MOLECULAR VIBRATIONS

An Algebraic and Nonlinear Approach

MOLECULAR VIBRATIONS

An Algebraic and
Nonlinear Approach

Guozhen Wu
Tsinghua University, China

Science Press

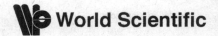
World Scientific

Published by

World Scientific Publishing Co. Pte. Ltd.

5 Toh Tuck Link, Singapore 596224

USA office: 27 Warren Street, Suite 401-402, Hackensack, NJ 07601

UK office: 57 Shelton Street, Covent Garden, London WC2H 9HE

British Library Cataloguing-in-Publication Data
A catalogue record for this book is available from the British Library.

MOLECULAR VIBRATIONS
An Algebraic and Nonlinear Approach

The Work was originally published by Science Press in 2014.
This edition is published by World Scientific Publishing Company
Pte Ltd by arrangement with Science Press, Beijing, China.

ISBN 978-981-3270-69-5

For any available supplementary material, please visit
https://www.worldscientific.com/worldscibooks/10.1142/10997#t=suppl

Desk Editor: Christopher Teo

Typeset by Stallion Press
Email: enquiries@stallionpress.com

Printed in Singapore

Contents

Preface   xi

1   Pendulum Dynamics   1

   1.1   Pendulum dynamics   1

   1.2   Morse oscillator   3

   1.3   Hamilton's equations of motion   6

   1.4   Pendulum dynamics as the basic unit

         for resonance   7

   1.5   Standard map and KAM theorem   9

   1.6   Conclusion 11

   References 11

2   Algebraic Approach to Vibrational Dynamics   13

   2.1   The algebraic Hamiltonian 13

   2.2   Heisenberg's correspondence and coset

         representation 15

   2.3   An example: The H_2O case 16

   2.4   su(2) dynamical properties 19

   Reference 22

   Appendix: The derivation of raising and lowering

           operators 23

3   Chaos   25

    3.1   Definition and Lyapunov exponent: Tent map . . .   25

    3.2   Lyapunov exponent in Hamiltonian system   28

    3.3   Period 3 route to chaos   28

    3.4   Resonance overlapping and sine circle mapping . . .   29

    3.5   The case study of DCN   32

        3.5.1   The chaotic motion   32

        3.5.2   Periodic trajectories   34

        3.5.3   Chaotic motion originating from the D–C
               stretching   42

    References   44

    Appendix: Calculation of the maximal Lyapunov
              exponent   44

4   C–H Bending Motion of Acetylene   47

    4.1   Introduction   47

    4.2   Empirical C–H bending Hamiltonian   48

    4.3   Second quantization representation of H_{eff}   49

    4.4   su(2) $\otimes$ su(2) represented C–H bending motion . . .   50

    4.5   Coset representation   52

    4.6   Modes of C–H bending motion   52

    4.7   Reduced Hamiltonian of C–H bending motion . . .   60

    4.8   su(2) origin of precessional mode   66

    4.9   Nonergodicity of C–H bending motion   68

    4.10   Intramolecular vibrational relaxation   74

    References   77

**5   Assignments and Classification of Vibrational
Manifolds   79**

    5.1   Formaldehyde case   79

    5.2   Diabatic correlation, formal quantum number
         and level reconstruction   81

    5.3   Acetylene case   85

    5.4   Background of diabatic correlation   88

    5.5   Approximately conserved quantum number   91

    5.6   DCN case   94

5.7 Density ρ in the coset space 98
5.8 Lyapunov exponent analysis 100
References 102

6 Dixon Dip **103**

6.1 Significance of level spacings 103
6.2 Dixon dip 103
6.3 Dixon dips in the systems of Henon–Heiles
and quartic potentials 104
6.4 Destruction of Dixon dip under multiple
resonances 106
6.5 Dixon dip and chaos 113
References 115

**7 Quantization by Lyapunov Exponent
and Periodic Trajectories** **117**

7.1 Introduction 117
7.2 Hamiltonian for one electron in multiple sites 118
7.3 Quantization: The least averaged Lyapunov
exponent 120
7.4 Quantization of H_2O vibration 123
7.5 Action integrals of periodic trajectories:
The DCN case 124
7.6 Retrieval of low quantal levels of DCN 128
7.7 Quantization of Henon-Heiles system 131
7.8 Quantal correspondence in the classical AKP
system 138
7.9 A comment 142
References 142

**8 Dynamics of DCO/HCO and Dynamical
Barrier Due to Extremely Irrational Couplings** **145**

8.1 The coset Hamiltonian of DCO 145
8.2 State dynamics of DCO 148
8.3 Contrast of the dynamical potentials of D–C
and C–O stretchings 152

8.4   The HCO case 155
8.5   Comparison of the dynamical potentials 157
8.6   A comment: The IVR role of bending motion 157
8.7   Dynamical barrier due to extremely irrational
       couplings: The role of bending motion 159
References 165

**9   Dynamical Potential Analysis for HCP, DCP,
       N$_2$O, HOCl and HOBr   167**

9.1   Introduction 167
9.2   The coset represented Hamiltonian of HCP 168
9.3   Dynamical potentials and state properties inferred
       by action population 170
9.4   State classification and quantal environments . . . 175
9.5   Localized bending mode 178
9.6   The condition for localized mode 182
9.7   On the HPC formation 182
9.8   The fixed point structure 183
9.9   DCP Hamiltonian 183
9.10   Dynamical similarity between DCP and HCP 188
9.11   N$_2$O dynamics 191
9.12   The cases of HOCl and HOBr 199
9.13   A comment 214
References 215
Appendix 215

**10   Chaos in the Transition State Induced
       by the Bending Motion   219**

10.1   Chaos in the transition state 219
10.2   The cases of HCN, HNC and the transition
       state 221
10.3   Lyapunov exponent analysis 224
10.4   Statistical analysis of the level spacing
       distribution 226
10.5   Dixon dip analysis 227

10.6 Coupling of pendulum and harmonic
        oscillator 227
10.7 A comment 230
References 230

**Appendix: Author's Publications
        Related to this Monograph   231**

Preface

The classical dynamical treatment in atomic physics, along with the ideas borrowed from the nonlinear dynamics and chaos theory, may not be unfamiliar to physicists. However, similar classical treatment in molecular spectroscopy/physics is barely popular. The traditional treatment in molecular spectroscopy/physics is based on the wave function algorithm. The topics of nonlinearity and chaos theory are hardly listed in the graduate programs in both the physics and chemistry departments. We have to admit this is a deficiency, if we can recognize that this classical algorithm is of high potential in the future development of molecular spectroscopy/physics.

Indeed, the concepts that are applicable in this field are not available at hand. Instead, it takes one's patience, imagination and elaboration to integrate these concepts altogether which are scattered around in various fields and can be new to one's background. In this aspect, learning from literatures is important. But this is not all. To learn these concepts through one's work is a necessity. It helps deepen the understanding of these concepts and stir up new ideas.

I am happy and lucky that in the past 20 years I had the chance to get involved in and acquainted with this field. My bias is that, not only is this classical treatment applicable to molecular vibrations but also it deepens our understanding of the molecular spectroscopy/physics. Moreover, it bridges the close relation between the classical and quantal dynamics.

This monograph compiles the highlights of my works in this field. These works have been published in the journals (as listed in Appendix) and partly in my monograph by Elsevier more than 12 years ago. Since then, more results have come out. They are re-organized now in this monograph in a more compact form.

The first three chapters deal with the main ideas that constitute the core concepts of this approach. Trying to have a compact presentation of these concepts that root in nonlinear dynamics and chaos theory, I did not try to plow to their very origins. Instead, only their meanings from a usage viewpoint are introduced. For those who are interested in their very origins, more readings shown in References and even in literatures are required. Nevertheless, it is stressed that the basic building bricks of a vibrational system, no matter how complicated it can be, are those comprised in the pendulum dynamics. The rest chapters cover their various applications in molecular vibrations. It includes: C–H bending motion of acetylene; Assignments and classification of vibrational manifolds; Dixon dip; Quantization by Lyapunov exponent and periodic trajectories; Dynamics of DCO/HCO, barrier due to extremely irrational couplings and the role of bending motion; Dynamical potential analysis for HCP, DCP, N_2O, HOCl and HOBr; Chaos in the transition state induced by the bending motion.

I hope the readers can appreciate, besides the power of this classical method, not only those results that may not be so transparent if the wave function approach is adopted but also the classical nature embedded in a quantal system. Classical and quantal aspects are complementary for the vibrational systems.

Of course, I am not the only contributor to these works. My former graduate students had also their roles. Over the past 20 years, the grant support by the China National Science Foundation and the State Key Laboratory of Low-dimensional Quantum Physics of the Department of Physics, Tsinghua University is greatly acknowledged.

Guozhen Wu
February 1, 2017
At Physics Department,
Tsinghua University, Beijing

Chapter 1

Pendulum Dynamics

1.1 Pendulum dynamics

The motion of a pendulum is well known. We will demonstrate throughout in this monograph that in fact, it plays an essential role in dynamics.

Shown in Fig. 1.1 is the configuration of a pendulum with coordinate, θ and its potential, V which is proportional to $1 - \cos\theta$. As $\theta = 0$, its potential is minimal. As $\theta = \pi$, it reaches the highest point and its potential is maximal. If its kinetic energy is not large enough, the pendulum will be bound in the region between $-\pi$ and π and its motion is periodic. In this situation, if the energy is small enough, that $1 - \cos\theta = 2\sin^2\theta/2$ can be approximated by $\theta^2/2$ (via $\sin\theta \approx \theta$), then the motion is the simple harmonic oscillator and the period is a constant, independent of the motion amplitude. Of course, for the motion with larger amplitude, the period will be energy dependent and the motion is called nonlinear or anharmonic. Since the potential valley is wider at its top, if the energy is quantized, the nearest neighboring level spacing will be smaller near the valley top. This is shown also in Fig. 1.1. As the pendulum energy is larger than the potential at $\theta = \pi$, the pendulum will rotate clockwise or counter-clockwise so the angle may be extended indefinitely. The clockwise and counter-clockwise rotations are equivalent with opposite actions. Shown in Fig. 1.2 is its phase diagram where J is

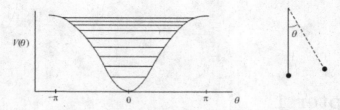

Fig. 1.1   The configuration of a pendulum with coordinate, θ and its potential, V which is proportional to $1 - \cos\theta$. The horizontal lines in the potential valley show the quantized levels. Due to the nonlinear effect, the nearest neighboring level spacing is less for higher levels.

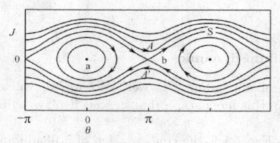

Fig. 1.2   The phase diagram of the motion of a pendulum. J is the action and θ is the angle. Curves correspond to the various quantized levels. Arrows denote the directions of motion. A, A' show the opposite rotations. S is the separatrix. a and b are the stable and unstable fixed points, respectively.

the rotational action (angular momentum). Figure 1.2 depicts the pendulum dynamics in a succinct way in which the periodic motions are the circles and the freely rotating motions are the up-and-down curves. The periodic motion is bound in a limited region while the rotating motion is unbound. These two motions are of different characters dynamically. They are separated by the curve labelled by S, called the separatrix. Besides, the two fixed points, a and b, deserve attention. Point a is dynamically stable while b is unstable where the pendulum is at its highest position. It deserves to note that point b is just the cross point of the two separatrix lines where the line (or flow line) directions are opposite: one is toward while the other is away from point b, just like that of a saddle (hyperbola). Hence, this unstable fixed point is also called the saddle point where along one direction, the dynamics is stable while along the other unstable.

We will see that in higher dimensions, the region near the saddle point is where chaotic motion will emerge first if it occurs. This was first visualized by Poincare who pointed out that once the stable and unstable orbits originating from the same saddle point intersect once, then they will intersect infinite times, forming the so-called homoclinic tangle. Thus, chaos forms around the saddle fixed point. The tangling orbit will approach the fixed point in an exponentially decelerating way and will never reach the fixed point.

This is a very important milestone as far as the developing of the chaotic dynamics is concerned. It deserves to quote what Poincare wrote in his *New Methods of Celestial Mechanics* in 1899: *When we try to represent the figures formed by these two curves and their infinitely many intersections, ... these intersections form a type of trellis, tissue or grid with infinitely fine mesh; neither of the two curves must ever cut across itself again, but it must bend back upon itself in a very complex manner in order to cut across all of the meshes in the grid an infinite number of times. The complexity of this figure is striking, and I shall not even try to draw it. Nothing is more suitable for providing us with an idea of the complex nature of the three body problem and of all the problems of dynamics in general.* Poincare found this phenomenon in his study of the three body problem. He also visualized that this chaotic issue is quite general.

We summarize the dynamical ideas embedded in the pendulum motion: (1) The harmonic and anharmonic periodic motions. The motion period of the former is energy (or action) independent while the latter is energy dependent. This difference can be used to define linearity and nonlinearity for a dynamical system. (2) The separatrix which divides the bound and unbound dynamical realms. (3) Fixed points which are either stable or unstable. (4) Chaotic motion which emerges around the unstable fixed point. All these form the basic features of a dynamical system.

1.2   Morse oscillator

Simple harmonic oscillator is often used to simulate a chemical bond. However, at high excitation, a bond may break down (or dissociation). Also, at high excitation, the nearest neighboring level spacing becomes narrower. These feathers cannot be modelled by simple harmonic oscillator. The commonly employed model is the Morse oscillator whose potential is

$$D\{1 - \exp[-a(r - r_0)]\}^2.$$

Here, a, D are constants, r_0 is the equilibrium position, $r - r_0$ ($\equiv \Delta r$) is the bond displacement and a^{-1} is the characteristic length. The Morse potential is shown in Fig. 1.3 where the horizontal line indicates the energy level with action n (or quantum number).

The eigenenergies of a Morse oscillator can be obtained by solving the Schroedinger equation. However, we will not follow this way. We suppose first that action n of a Morse oscillator can be written as $(1/2)$ $(q^2 + p^2)$. (q, p) are the generalized coordinates. (In the next chapter, it will be clear that they are the so-called coset coordinates). For convenience, we may consider q as the quantity related to the displacement Δr and p the quantity to the momentum of the oscillator. Then as $p = 0$, the displacement reaches its extrema, denoted as Δr_0^+ and Δr_0^-. Their relations with q is shown by the following formula[1]

$$\Delta r_0 = a^{-1} \, ln \, \{[1 - (1 - \lambda^2)^{1/2}\mathrm{sgn}]/\lambda^2\}. \qquad (1.1)$$

Here, as $q < 0$, sgn $= 1$, Δr_0 is Δr_0^-; as $q > 0$, sgn $= -1$, Δr_0 is Δr_0^+. Meanwhile, $\lambda = 1 - (2n + 1)/k$ and $2n = q^2$. k is a parameter

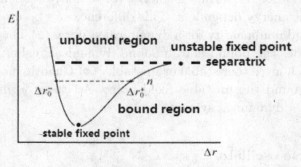

Fig. 1.3   The Morse potential. Δr_0^+ and Δr_0^- show the displacement extrema for a level (the horizontal line) with action n. Also shown are the dynamical ingredients.

whose physical meaning will be given later. From (1.1), we have:

$$1 - e^{-a\Delta r_0} = -(1 - \lambda^2)^{1/2} \quad \mathrm{sgn} = 1,$$
$$= (1 - \lambda^2)^{1/2} \quad \mathrm{sgn} = -1$$

or

$$(1 - e^{-a\Delta r_0})^2 = 1 - \lambda^2.$$

This means that Morse potential is $D(1 - \lambda^2)$.

Since $\lambda = 1 - (2n + 1)/k$, we have

$$D(1 - \lambda^2) = (4D/k)(n + 1/2) - (4D/k^2)(n + 1/2)^2. \tag{1.2}$$

Note that $D(1 + \lambda)(1 - \lambda) = D(1 - \lambda^2)$ is the potential at the extremal displacements. At these displacements, the kinetic energy is zero. Hence, (1.2) is the system energy and under quantization, it is the eigenenergy for the level with quantum number n. The right-hand-side expression shows that this is a second order anharmonic oscillator with $-4D/k^2$ the nonlinear coefficient and $4D/k$ the linear coefficient for the harmonic part. It is now clear that the Morse oscillator is a system of second order anharmonicity. In spectroscopy, we generally denote the eigenenergy E_n of such a system as

$$\omega(n + 1/2) + X(n + 1/2)^2. \tag{1.3}$$

Hence, for a Morse oscillator:

$$\omega = 4D/k, \quad X = -4D/k^2$$

or

$$k = \omega/(-X), \quad D = (\omega/2)^2/(-X).$$

The significance is that we now relate the Morse parameters k, D to the spectroscopic parameters ω, X.

The energy spacing between two nearest neighboring levels becomes smaller as n is larger. As the spacing is zero, the oscillator dissociates. We denote the quantum number as n_0 at dissociation.

From

$$E_{n_0+1} - E_{n_0} = 0$$

we have

$$\omega + X(2n_0 + 2) = 0$$

or

$$n_0 = (1/2)(\omega/(-X)) - 1 = (k/2) - 1 \approx k/2.$$

Hence, k is twice the dissociation quantum number which can be obtained from the spectroscopic parameters, ω and X.

Finally, we note that $a = \omega\sqrt{\mu/2D} = \sqrt{-2X\mu}$ with μ the reduced mass of the Morse oscillator.

Substituting $n_0 = (1/2)(\omega/(-X)) - 1$ into (1.3), we obtain that the dissociation energy is $(\omega^2 - X^2)/(-4X) \sim D$. For instance, for D—CN bond, from the experiment, we have $\omega = 2681.4\,\text{cm}^{-1}$, $X = -21.2\,\text{cm}^{-1}$, then n_0 is 63 and D is $84761\,\text{cm}^{-1}$.

In conclusion, we note that Morse oscillator is the *simplest* model other than the simple harmonic oscillator that can simulate bond vibration including its dissociation. This is the reason that the Morse model is prevalently employed in molecular spectroscopy.

Dynamically, as shown in Fig. 1.3, we note that the bottom of the Morse potential valley is a stable fixed point. In the Morse valley, the motion is periodic and anharmonic. Of course, close to the very valley bottom, the motion can be a simple harmonic one. The separatrix is the bond dissociation. Above dissociation, the motion is unbound. If chaos appears, it will be during the bond dissociation. We also note that near the separatrix, the nearest neighboring level spacing reaches its minimum.

1.3   Hamilton's equations of motion

In Hamilton-Jacobi mechanics, the system Hamiltonian H is expressed in terms of action-angle variables, $(J,\ \theta)$ and the equations of motion are the following equation set:

$$\partial H/\partial J = \dot{\theta} = \omega, \quad \partial H/\partial\theta = -\dot{J}.$$

The upper dot sign denotes time derivative. The action-angle variables can be generalized coordinate pairs, (q, p). This formulation will be convenient for our study.

The Hamiltonian of a pendulum can be cast as

$$Gp^2/2 - F\cos\theta$$

with constant G and F and the equations of motion are:

$$\dot{p} = -F\sin\theta, \quad \dot{\theta} = Gp$$

1.4 Pendulum dynamics as the basic unit for resonance

Since vibrational action is quantized, vibrational energy transfer between two bonds in a molecule has to satisfy the condition: $\omega_1 \approx \omega_2$, for instance for the two O-H bonds in H_2O. The energy transfer between bonds is a resonance effect.

Suppose $H_0(I_1, I_2)$ is the Hamiltonian for a system of two independent vibrating bonds. I_1, I_2 are their actions. Then we have: $\omega_1^0 = \partial H_0/\partial I_1$, $\omega_2^0 = \partial H_0/\partial I_2$ and $\omega_1^0 \approx \omega_2^0$.

The Hamiltonian under resonance can be:

$$H = H_0(I_1, I_2) + C_0(I_1, I_2)\cos(\varphi_1 - \varphi_2)$$

where C_0 is the coupling strength, φ_1 and φ_2 are the respective angles. Under the transformations:

$$J_1 = (I_1 - I_2)/2, \quad \theta_1 = (\varphi_1 - \varphi_2)$$
$$J_2 = (I_1 + I_2)/2, \quad \theta_2 = (\varphi_1 + \varphi_2)$$

then

$$H = H_0(J_1, J_2) + C_0(J_1, J_2)\cos\theta_1.$$

Since H is independent of θ_2, from $\partial H/\partial\theta_2 = -\dot{J}_2 = 0$, J_2 is a constant.

Then

$$H \approx (\partial H_0/\partial J_1)_0 (J_1 - J_1^0)$$
$$+ \frac{1}{2}(\partial^2 H_0/\partial J_1^2)_0 (J_1 - J_1^0)^2 + C_0(J_1^0, J_2^0)\cos\theta_1$$

but

$$(\partial H_0/\partial J_1)_0 = \frac{\partial H_0}{\partial I_1}\frac{\partial I_1}{\partial J_1} + \frac{\partial H_0}{\partial I_2}\frac{\partial I_2}{\partial J_1} = \omega_1^0 - \omega_2^0 = 0$$

hence,

$$H \sim \alpha J_1^2/2 + C_0 \cos\theta_1$$

with α, C_0 constants. This is exactly the Hamiltonian of a pendulum in a potential $C_0 \cos\theta_1$ with kinetic energy $\frac{1}{2}\alpha J_1^2$. Therefore, under resonance, locally the phase structure is approximated by the pendulum dynamics as shown in Fig. 1.2. Physically, this consequence is very important. It means that under a resonance, the dynamics is just that of a pendulum. Explicitly, a resonance will induce these dynamical ingredients: harmonic and anharmonic periodic motions, separatrix, stable and unstable fixed points and chaotic motion. All these dynamical ingredients are *coexistent* in the system.

Based on Fig. 1.2, the vibrational dynamics is interpreted as follows. The two fixed points correspond to that the actions on the two bonds are the same and they are in phase and out of phase, respectively. This is the mode-locking. For small J_1 the action transfer between the bonds is strong (J_1 changes it sign). The coupling is strong that their phase difference is restricted only in a certain range. The mode is called *normal*. For larger J_1 above (or lower than) the separatrix, J_1 does not change sign. The coupling is weaker. Then the action is more or less aggregated in either bond. The phase difference between the two vibrating bonds is rather irrelevant. It can extend from 0 to 2π. The mode is called *local*. Chaos will emerge in between the local and normal modes where the separatrix stays. The two modes, corresponding to the clockwise and counter-clockwise pendulum motions, are doubly degenerate.

1.5 Standard map and KAM theorem[2]

The Hamiltonian of a rotor with rotational inertia, I under periodic impulses of period τ as shown in Fig. 1.4 is:

$$H(p_\theta, \theta, t) = \frac{p_\theta^2}{2I} + k\cos\theta \sum_m \delta(t - m\tau).$$

Here, p_θ is angular momentum. δ is the Dirac function.

The equations of motion are:

$$\dot{p}_\theta = k\sin\theta \sum_m \delta(t - m\tau)$$

$$\dot{\theta} = p_\theta/I.$$

Between two consecutive impulses, p_θ is a constant and θ increases monotonically. After each impulse, the change of p_θ is:

$$\int_{t=n\tau+0^+}^{t=n\tau+1+0^+} dp_\theta = \int k\sin\theta \sum_m \delta(t - m\tau)dt = k\sin\theta_{n+1}.$$

Hence, we have:

$$p_{n+1} = p_n + k\sin\theta_{n+1}$$

$$\theta_{n+1} = \theta_n + p_n\tau/I.$$

(For convenience, we may let $\tau/I = 1$)

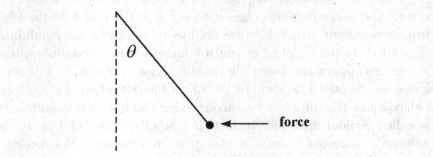

Fig. 1.4   The rotor under periodic impulses.

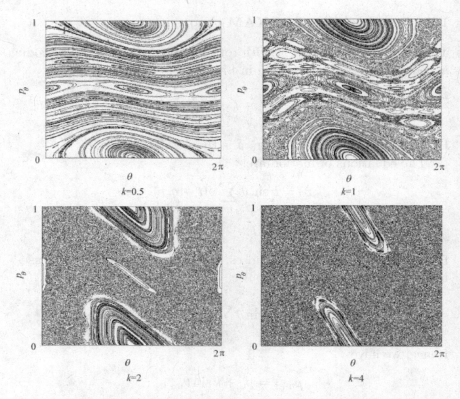

Fig. 1.5   Standard map with various coupling strengths.

This is a very simple map and is often called *standard map*. Figure 1.5 shows the mapping cases with $k = 0.5$, 1, 2 and 4. In the Figure, we see many structures that are like those due to the pendulum dynamics. As the coupling strength k increases, more pendulum-like structures appear and more the chaotic regions prevail. As k is very large, the chaotic region will be global. In such situation, a trajectory behaves like the diffusion by encountering the hyperbolic points. It is called Arnold diffusion. In fact, if a small portion of Fig. 1.5 is enlarged, more pendulum-like structures will appear. This process shows self-similarity.

Mapping is an important idea and is useful in dynamical study. Computing simulation is basically by mapping. Now we put the

Hamilton's equations of motion for pendulum in the form of mapping:

$$p_{n+1} - p_n = -F \sin \theta_n$$

$$\theta_{n+1} - \theta_n = Gp_{n+1}.$$

Apparently, this is identical to the standard map by rescaling.

The way the birth of pendulum-like structures in a dynamical system is an essential topic. This is addressed by the so-called KAM (by Kolmogorov, Arnold and Moser) theorem. The theorem is rather complicated, but it states basically that:

For a system of N oscillators of periods (or angular frequencies) $\{\omega_{01}, \ldots, \omega_{0N}\}$, as there are interactions among them, the trajectories, of which $\{\omega_{01}, \ldots, \omega_{0N}\}$ are in rational ratios, will be destroyed first and bifurcate into pendulum-like structures with the birth of chaotic trajectories, separatrices, stable and unstable fixed points, and periodic trajectories with new periods that are in rational or irrational ratios. (For irrational ratio case, the motion is called quasiperiodic since it is not exactly periodic.) While those trajectories with irrational ratios of $\{\omega_{01}, \ldots, \omega_{0N}\}$ will persist till the interaction is stronger. The newly born periodic trajectories will be destroyed in a similar way with the birth of more chaotic and periodic trajectories along with separatrices and fixed points. This is an infinite self-similar process.

1.6   Conclusion

The core idea of this chapter is that the pendulum dynamics plays the basic role in classical mechanics. Interaction which is the central issue in physics leads to the pendulum dynamics. Pendulum dynamics possesses these elements: chaos, stable and unstable fixed points, periodic motion, bound and unbound regions, separatrix. They form the dynamical key ingredients.

References

1. Rankin C C, Miller W H. *J. Chem. Phys.*, 1971, 55: 315.
2. Ott E. Chaos in Dynamical Systems. Cambridge: Cambridge University Press, 1994.

Chapter 2

Algebraic Approach to Vibrational Dynamics

2.1 The algebraic Hamiltonian

Action-angle variables are employed in classical Hamilton-Jacobi mechanics. In quantum mechanics, action is quantized in the unit of Planck's constant, $h/2\pi$. Interaction or resonance is the transfer of actions among the bond vibrations. This can be realized by the second quantization operators, a^+ and a which are the raising and lowering operators possessing the following properties. (An appendix is attached in the end of this chapter for those who need their derivation.)

$$a^+|n\rangle = \sqrt{n+1}|n+1\rangle$$
$$a|n\rangle = \sqrt{n}|n-1\rangle.$$

Here, n is an integer. For instance, for the resonance between two equivalent bonds, like the two O-H stretchings in H_2O, the coupling can be $a_s^+ a_t + h.c.$ where s and t denote the two stretching coordinates. ($h.c.$ stands for hermitian conjugate) This is usually called 1:1 resonance. Sometimes, the second order resonance, or the Darling-Dennison coupling is considered with the form: $(a_s^+ a_s^+ a_t a_t + h.c.)$. This is the 2:2 resonance. For the non-resonant Hamiltonian part,

13

that of the Morse oscillator can be employed. So the complete algebraic Hamiltonian for two interacting equivalent bonds is

$$\omega_s(n_s + 1/2) + \omega_s(n_t + 1/2) + X_s(n_s + 1/2)^2$$
$$+ X_s(n_t + 1/2)^2 + X_{st}(n_s + 1/2)(n_t + 1/2)$$
$$+ K_{st}(a_s^+ a_t + h.c.) + K_{DD}(a_s^+ a_s^+ a_t a_t + h.c.).$$

Very often, the Fermi resonance between the bond stretching and bending is operative (the energy of two bending actions is roughly the same as one stretching action). The coupling is 1:2 resonance with the form, $a_s^+ a_b a_b + a_t^+ a_b a_b + h.c.$ If the Fermi resonance is included, the Hamiltonian is

$$\omega_s(n_s + 1/2) + \omega_s(n_t + 1/2) + \omega_b(n_b + 1/2)$$
$$+ X_s(n_s + 1/2)^2 + X_s(n_t + 1/2)^2 + X_b(n_b + 1/2)^2$$
$$+ X_{st}(n_s + 1/2)(n_t + 1/2) + X_{sb}(n_s + n_t + 1)(n_b + 1/2)$$
$$+ K_{st}(a_s^+ a_t + h.c.) + K_{DD}(a_s^+ a_s^+ a_t a_t + h.c.)$$
$$+ K_F(a_s^+ a_b a_b + a_t^+ a_b a_b + h.c.).$$

It is interesting to note that for these two Hamiltonians, besides that the system energy is conserved, the sums of the total actions $n_s + n_t$ and $n_s + n_t + n_b/2$ are conserved, respectively, even the individual actions, n_s and n_t, are *destroyed* and not conserved due to couplings. The conserved action sum is called the *polyad number* (denoted by P) which is an important constant of motion.

Therefore, if the coefficients of the Hamiltonian are known, one may construct the Hamiltonian matrix under the bases $\{|n_s, n_t\rangle\}$ with integers n_s and n_t. Since the matrix is block diagonalized by the polyad numbers, we can simply choose a polyad number and then diagonalize the corresponding matrix to obtain the eigenenergies. In fact, we can reverse this process by fitting the calculated eigenenergies to the experimental values to elucidate the coefficients, ω, X and K, in the algebraic Hamiltonian. In this way, the algebraic Hamiltonian is closely based on the experimental data.

2.2 Heisenberg's correspondence and coset representation

Heisenberg proposed that the second quantization operators can be related to classical action n and angle ϕ as:

$$a^+ \approx \sqrt{n}e^{i\phi}, \quad a \approx \sqrt{n}e^{-i\phi}.$$

Meanwhile, generalized coordinates can be chosen as:

$$q_\alpha = \sqrt{2n_\alpha}\cos\phi_\alpha, \quad p_\alpha = -\sqrt{2n_\alpha}\sin\phi_\alpha \ (\alpha = s, t),$$

then

$$a_s^+ a_t + a_t^+ a_s \approx q_s q_t + p_s p_t.$$

(Note other choices are also possible like $q_\alpha = \sqrt{n_\alpha/2}\cos\phi_\alpha$, $p_\alpha = -\sqrt{n_\alpha/2}\sin\phi_\alpha$), $\phi_\alpha = \tan^{-1}(-p_\alpha/q_\alpha)$

This transformation bears its mathematical foundation on the group theory (such as the Lie group and Lie algebra), to which we will not go that far for the moment. The (q, p) coordinates are those for the so-called coset space. These expressions are called in the coset representation.

For the above mentioned interaction, since the total action $n_s + n_t$ is conserved, we can also adopt

$$a_s^+ \approx \sqrt{n_s}, \quad a_t^+ \approx \sqrt{n_t}e^{i\phi_t}.$$

Here, ϕ_t is the phase difference between s and t actions. Thereby, we have

$$a_s^+ a_t + a_t^+ a_s \approx \sqrt{2n_s}q_t.$$

Listed below are the two representations for convenient choice:

$$(1) \ n_s = P - n_t - n_b/2, \quad n_t = (q_t^2 + p_t^2)/2,$$
$$n_b = (q_b^2 + p_b^2)/2$$

and

$$K_{st}(2n_s)^{\frac{1}{2}}q_t$$

$$K_{DD}n_s(q_t^2 - p_t^2)$$

$$K_F\{\sqrt{n_s}(q_b^2 - p_b^2) + [q_t(q_b^2 - p_b^2) + 2p_t q_b p_b]/\sqrt{2}\}$$

$$(2) \quad n_b = 2(P - n_s - n_t), \quad n_s = (q_s^2 + p_s^2)/2,$$

$$n_t = (q_t^2 + p_t^2)/2,$$

and

$$K_{st}(q_s q_t + p_s p_t)$$

$$K_{DD}\left[\frac{1}{2}(q_s^2 - p_s^2)(q_t^2 - p_t^2) + 2q_s q_t p_s p_t\right]$$

$$K_F\sqrt{2}n_b(q_s + q_t).$$

In the coset representation, Hamilton's equations are:

$$\partial H/\partial p_\alpha = dq_\alpha/dt, \quad \partial H/\partial q_\alpha = -dp_\alpha/dt.$$

2.3   An example: The H_2O case

The two coupled O-H stretchings of H_2O are taken as an example. The system can be considered as the motion of two coupled pendula. Figure 2.1 shows its potential curve in which the broken line corresponds to the separatrix. a, a' are the stable fixed points, b is the unstable fixed point. For the levels above the separatrix, due to strong coupling, the motion will be normal and the energy spacing between

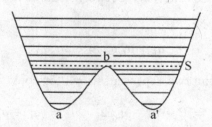

Fig. 2.1   The vibrational potential of two coupled stretchings. The horizontal lines are the energy levels. a, a' are the stable fixed points. b is the unstable fixed point. S is the separatrix.

the nearest neighboring levels becomes larger. The vibrational modes correspond to a, a' are the local modes with weak coupling for which the phase angles of the two oscillations are more or less independent without fixed relationship.

The Hamiltonian is put in a simple form

$$H = \omega_s(n_s + 1/2) + \omega_s(n_t + 1/2) + X_s\{(n_s + 1/2)^2 + (n_t + 1/2)^2\}$$
$$+ X_{st}(n_s + 1/2)(n_t + 1/2) + K_{st}(a_s^+ a_t + a_t^+ a_s).$$

The corresponding Hamiltonian in the coset representation is

$$\omega_s(P + 1) + \omega_s(n_t + 1/2) + X_s\{(n_s + 1/2)^2 + (n_t + 1/2)^2\}$$
$$+ X_{st}(n_s + 1/2)(n_t + 1/2) + 2K_{st}(n_s n_t)^{1/2} cos\varphi_{st},$$

where φ_{st} is the phase difference. $P = n_s + n_t$ and $n_s = P - n_t$.

As P is given, the Hamiltonian matrix is constructed from the bases: $\{|n_s, n_t\rangle = |0, P\rangle, |1, P - 1\rangle, \ldots, |P - 1, 1\rangle, |P, 0\rangle\}$. There are $P+1$ bases and there are $P+1$ eigenenergies. If we set the eigenenergy equal to the Hamiltonian, then we can obtain the relations of $n_s = n_s(\varphi_{st})$ and $n_t = n_t(\varphi_{st})$. This is the phase diagram.

The coefficients of the Hamiltonian were elucidated from the fit to the experimental values. They are (in cm^{-1}) : $\omega_s = 3890.6$, $X_s = -82.1$, $X_{st} = -13.2$ and $K_{st} = -42.7$.

Figure 2.2(a) is the phase diagram $((n_s - n_t)$ against $\varphi_{st})$ for the case of $P = 10$ (totally 11 levels). It is observed that for the low levels with small coupling, the two O-H stretchings possess quite the same energy and the phase angles are more or less independent that their difference, φ_{st}, can cover the whole range of $(-\pi, \pi)$. This shows the character of a local mode. At higher excitation, the coupling is stronger and the nearest neighboring level spacings become wider. Meanwhile, the phase angle difference is fixed within an interval around π. This is the character of an anti-symmetric normal mode. Figure 2.2(a) shows that the separatrix where the transition between the local and normal modes occurs is in between the 8th and 9th levels. Figure 2.2(b) shows that the energy spacing ΔE between the nearest neighboring levels (excluding the spacings of the levels that are approximately degenerate) is the smallest around the separatrix. This is consistent with the Morse (pendulum) case as mentioned in

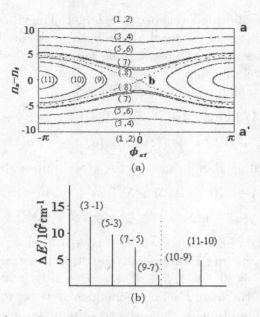

Fig. 2.2   The phase diagram of the vibrational system of two coupled O-H stretchings (a) and the nearest neighboring level spacings (b). The number in the parenthesis shows the level numbering. The dashed line shows the separatrix.

Chapter 1. We will see in the coming chapters that this is a generic dynamical phenomenon called Dixon dip. It is interesting to note that for levels lying above the separatrix, their spacing becomes wider. Apparently, they lie in an *anti-Morse potential* (or inverse Morse potential) where the highest point is a stable fixed point. This is schematically shown in Fig. 2.3.

If more couplings are involved than this system, we may infer that the motion for the levels close to the separatrix is more chaotic while that for the higher or lower levels is more regular. At a first glance, this seems contradicting to our common sense that higher energy is more chaotic. The reason is that at high excitation, stronger coupling will lead to the phase-locking phenomenon and the motion will tend to be more regular.

In conclusion, by the classical algebraic Hamiltonian, the dynamical characters of the highly excited vibration of two coupled oscillators can be retrieved. They are in consistency with those predicted by the pendulum dynamics as shown in Chapter 1.

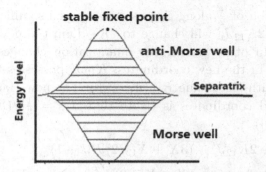

Fig. 2.3   The levels associated with a polyad number behave like staying in a Morse well for those below the separatrix and in an anti-Morse well for those above the separatrix.

Up to this point, we introduced the important dynamical ideas: (1) a resonance is accompanied by a polyad number which is a constant of motion, (2) the levels above the separatrix lie in an anti-Morse potential where the highest point is a stable fixed point, and (3) the Dixon dip near the separatrix where the nearest neighboring level spacing reaches its minimum.

2.4   su(2) dynamical properties[1]

For two equivalent Morse stretchings, consider only the 1:1 resonance, the Hamiltonian can be

$$\omega_0(n_1 + n_2 + 1) + X\left[\left(n_1 + \frac{1}{2}\right)^2 + \left(n_2 + \frac{1}{2}\right)^2\right]$$

$$+ X_{12}\left(n_1 + \frac{1}{2}\right)\left(n_2 + \frac{1}{2}\right) + K_{12}(a_1^+ a_2 + a_2^+ a_1).$$

We define $J_+ = a_1^+ a_2$, $J_- = a_2^+ a_1$, $J_z = \frac{1}{2}(a_1^+ a_1 - a_2^+ a_2)$ (or $J_x = (J_+ + J_-)/2$, $J_y = -\frac{i}{2}(J_+ - J_-)$). It can be shown that these three operators form an su(2) algebra just like the angular moment operators. Then, the Hamiltonian can be written as:

$$\omega_0(2J + 1) + 2K_{12}J_x + (2X + X_{12})J(J + 1)$$

$$+ (2X + X_{12})/4 + (2X - X_{12})J_z^2$$

with $2J = n_1 + n_2$ which is a constant.

The rotation of $\frac{\pi}{2}$ along J_y possesses special significance. This is because that $2K_{12}J_x$ will change to $2K_{12}J'_z$ in the new coordinates $\{J'_x, J'_y, J'_z\}$. In other words, the quantization changes from along J_z to $J_x(J'_z)$. In the new coordinates, $K_{12}J'_z$ possesses the eigenvector $|n'_1, n'_2\rangle$ with eigenvalue $K_{12}(n'_1 - n'_2)$. The new Hamiltonian H' in the rotated coordinates is (Note that $J'^2_x = \frac{1}{2}[J(J+1) - J'^2_z] + \frac{1}{4}(J'^2_+ + J'^2_-)$)

$$\omega_0(2J+1) + 2K_{12}J'_z + (3X + X_{12}/2)J(J+1)$$
$$+ (2X + X_{12})/4 - (X - X_{12}/2)J'^2_z + [(2X - X_{12})/4](J'^2_+ + J'^2_-).$$

Before rotation, the two Morse oscillators are identical, we call the coordinates the local mode picture. After rotation, in H', the two Morse oscillators are no longer identical. We call the rotated coordinates the normal mode picture (In fact, n'_1, n'_2 are the actions of the symmetric and antisymmetric modes, respectively. We note also that $J = \frac{1}{2}(n_1 + n_2) = \frac{1}{2}(n'_1 + n'_2)$). The rotated Hamiltonian now contains the Darling–Dennison term.

For these two pictures, we have (q, p) and (q', p'). Their relation can be determined by:

$$J'_x = -J_z, \quad J'_y = J_y, \quad J'_z = J_x.$$

Explicitly,

$$2q'[J - (q'^2 + p'^2)]^{\frac{1}{2}} = J - 2(q^2 + p^2) \tag{2.1}$$

$$-2p'[J - (q'^2 + p'^2)]^{\frac{1}{2}} = -2p[J - (q^2 + p^2)]^{\frac{1}{2}} \tag{2.2}$$

$$2(q'^2 + p'^2) - J = 2q[J - (q^2 + p^2)]^{\frac{1}{2}}. \tag{2.3}$$

Dividing (2.2) by (2.1), we have

$$p' = \frac{2p[J - (q^2 + p^2)]^{\frac{1}{2}}}{J - 2(q^2 + p^2)} q'. \tag{2.4}$$

By substituting (2.4) into (2.3), we have $q' = q'(q, p)$ Explicitly, it is:

$$q' = \text{sgn} \bullet \left[\left(q[J - (q^2 + p^2)]^{\frac{1}{2}} + \frac{J}{2} \right) \left(1 + \frac{4p^2[J - (q^2 + p^2)]}{[J - 2(q^2 + p^2)]^2} \right)^{-1} \right]^{\frac{1}{2}}.$$
$$\tag{2.5}$$

Here,

$$\text{sgn} = +1, \quad \text{if } J - 2(q^2 + p^2) > 0;$$
$$\text{sgn} = -1, \quad \text{if } J - 2(q^2 + p^2) < 0;$$

as

$$J = 2(q^2 + p^2), \quad q' = p' = 0.$$

From (2.4) and (2.5), we have $p' = p'(q,p)$. (2.4) and (2.5) are very useful. As soon as the dynamics in one picture is known, then the dynamics in another picture can be known immediately via the transformation between (q,p) and (q',p').

For an eigenstate, we may equate its energy E to $H(q,p)$ or $H'(q',p')$ by which $q = q(p)$ or $q' = q'(p')$ can be obtained. Sometimes, the relation (J_z, ϕ) or (J'_z, ϕ') can be more useful. They show the relation of the action difference versus the angle difference in the local or normal mode picture.

We take H_2O and O_3 as examples. Given the polyad number $P = n_1 + n_2$, there are $P + 1$ levels. They are designated as L# with # = 1, 2, 3, ... Levels are designated from the lowest to the highest.

There are two classes of levels. One is with two concentric trajectories in (q,p) coordinates, which show up as two closed trajectories in (q',p') coordinates. In (J_z, ϕ) and (J'_z, ϕ') coordinates, they are non-localized (ϕ extends from 0 to 2π) and localized (ϕ' is limited in a certain range), respectively. This is the character of the local mode for which the two oscillators are weakly coupled with no fixed phase difference. The two separated trajectories in (J_z, ϕ) coordinates correspond to the clockwise and counter-clockwise motions of the pendulum. Note that these trajectories in the normal mode picture are *localized* ! Their ϕ' are limited to 0 and π. Figure 2.4 shows the character of the local mode for the case of O_3, $P = 8$, L3 as an example.

Another class is with closed trajectories in both (q,p) and (q',p') coordinates. In (J_z, ϕ) coordinates, ϕ is limited to around π and in (J'_z, ϕ') coordinates, ϕ' is non-localized. This class is the normal mode. The localization of ϕ shows that the coupling between the two bond vibrations is strong enough for them to possess a relatively fixed

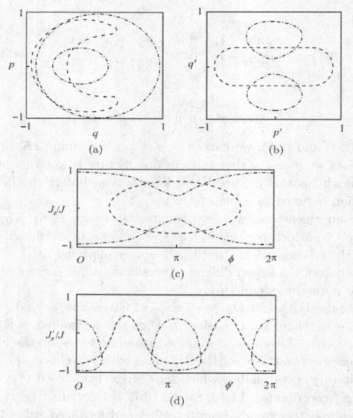

Fig. 2.4   The trajectories of O_3, $P = 8$, L3 (–·–·–·–) and H_2O, $P = 4$, L3 (.......).

phase relation. That ϕ is around π shows that it is antisymmetric. (As the coupling is even stronger, for higher levels, ϕ can be limited to 0 and it is a symmetric normal mode.) Figure 2.4 shows the character of the normal mode for the case of H_2O, $P = 4$, L3 as an example.

In conclusion, for a given P, there are $P + 1$ levels. Those with lower energy are local modes while those with higher energy are normal. There is the local/normal transition

Reference

1. Kellman M E. *J. Chem. Phys.*, 1985, 83: 3843.

Appendix: The derivation of raising and lowering operators

We write the Hamiltonian for a simple harmonic oscillator as:

$$H = \frac{1}{2}(p^2 + \omega^2 q^2)$$

p, q are the conjugate momentum and coordinate. They satisfy the quantization condition:

$$[q, p] \equiv qp - pq = i\hbar.$$

Then, we have

$$[H, p] = i\hbar\omega^2 q, \quad [H, q] = -i\hbar p.$$

Suppose the eigen equation $H|n\rangle = E_n|n\rangle$ and define

$$a^+ = -ip + \omega q,$$

then

$$
\begin{aligned}
Ha^+|n\rangle &= H(-ip + \omega q)|n\rangle \\
&= \{-i[H, p] - ipH + \omega[H, q] + \omega qH\}|n\rangle \\
&= \{-i(i\hbar\omega^2 q) - ipE_n + \omega(-i\hbar p) + \omega qE_n\}|n\rangle \\
&= (E_n + \hbar\omega)a^+|n\rangle.
\end{aligned}
$$

Hence, $a^+|n\rangle$ possess an extra $\hbar\omega$ than $|n\rangle$. Similarly, for $a = ip + \omega q$, we have

$$Ha|n\rangle = (E_n - \hbar\omega)a|n\rangle$$

i.e., $a|n\rangle$ possess a less $\hbar\omega$ than $|n\rangle$.

For the ground state $|0\rangle$, $a|0\rangle = 0$

$$
\begin{aligned}
a^+a|0\rangle &= (p^2 + \omega^2 q^2 + i\omega[q, p])|0\rangle \\
&= (2H - \hbar\omega)|0\rangle \\
&= (2E_0 - \hbar\omega)|0\rangle = 0
\end{aligned}
$$

then

$$E_0 = \frac{1}{2}\hbar\omega.$$

Therefore the eigenenergy for $|n\rangle$ is:

$$E_n = \left(n + \frac{1}{2}\right)\hbar\omega, \quad n = 0, 1, 2, \ldots,$$

The results are:

$$H|n\rangle = \left(n + \frac{1}{2}\right)\hbar\omega|n\rangle$$

$$a^+|n\rangle \sim |n+1\rangle$$

$$a|n\rangle \sim |n-1\rangle.$$

Now we require

$$a^+|n\rangle = \sqrt{n+1}|n+1\rangle$$

$$a|n\rangle = \sqrt{n}|n-1\rangle$$

and put $a^+ = A(-ip + \omega q)$, $a = A(ip + \omega q)$
By

$$aa^+|n\rangle = (n+1)|n\rangle$$
$$= A^2(ip + \omega q)(-ip + \omega q)|n\rangle$$
$$= A^2\{2H + i\omega[p, q]\}|n\rangle$$
$$= A^2 2(n+1)\hbar\omega|n\rangle$$

A is obtained as $A = 1/\sqrt{2\hbar\omega}$

Chapter 3

Chaos

3.1 Definition and Lyapunov exponent: Tent map

For a forced damped pendulum, the differential equation of motion is:

$$\ddot{\theta} + \nu\dot{\theta} + \sin\theta = T\sin\omega t.$$

The $\dot{\theta}$ term is for the damping, $\sin\omega t$ is the forced term. Let $x_1 = \dot{\theta}$, $x_2 = \theta$, $x_3 = \omega t$, then

$$\dot{x}_1 = T\sin x_3 - \sin x_2 - \nu x_1$$
$$\dot{x}_2 = x_1$$
$$\dot{x}_3 = \omega.$$

This kind of equation set is called N-order autonomous differential equations. For this case $N = 3$. This kind of mapping is often encountered in dynamical study. Theorem tells that as $N \geq 3$, trajectories can be chaotic while for $N = 1, 2$, there will be no chaotic trajectories. For the forced damped pendulum, $N = 3$, hence, its motion can be chaotic. For a map, if it is invertible and with $N \geq 2$, then its trajectories can be chaotic. Otherwise, if $N = 1$, the motion is regular and ordered. If a map is noninvertible, then even with $N = 1$, the motion can be chaotic.

Chaotic motion originates from the intertwining of two kinds of mapping: augmentation and compression. This leads to that its trajectories are very initial point dependent. Suppose initially, two points have a deviation $\Delta(0)$ and at t, the deviation is $\Delta(t)$. If

$$\Delta(t) \approx \Delta(0)e^{ht}$$

and $h > 0$, then the trajectories are very initial point dependent. h, called the *Lyapunov exponent*, is a parameter for describing the *degree* of chaos. Now, it is accepted as the definition of chaos: a trajectory is chaotic if it follows a deterministic equation of motion and its Lyapunov exponent is larger than 0.

As an example, for a binary shift map, error is augmented 2^n folds after n maps. Since $2^n = e^{n \ln 2}$, $h = \ln 2$.

Chaotic motion may appear even in very simple systems. 1-dimensional tent map is employed for demonstration, which is shown in Fig. 3.1.

This transformation is to map the points in [0,1] according to the rules:

$$\text{As } X_n < 1/2, \quad X_{n+1} = 2X_n$$
$$\text{As } X_n > 1/2, \quad X_{n+1} = 2(1 - X_n).$$

This transformation augments 2-fold for the number less than 1/2 while it compresses the number greater than 1/2. Here, we see the augmentation and compression. Hence, chaos in this system is expected.

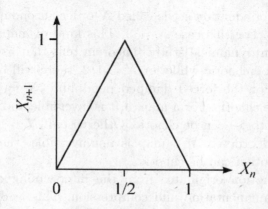

Fig. 3.1   The tent map.

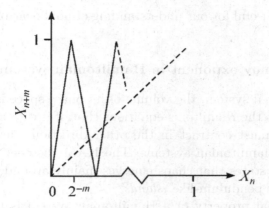

Fig. 3.2   The map from X_n to X_{n+m}. The dash line is the diagonal.

After m mappings, the relation between X_n and X_{n+m} is as shown in Fig. 3.2:

The significance of Fig. 3.2 is that two points within a deviation of 2^{-m} will be scattered around *chaotically* anywhere in $[0, 1]$ after m mappings no matter how small 2^{-m} is. The Lyapunov exponent is ln 2.

In $[0, 1]$, except chaotic points, there are numerous periodic points. As shown in Fig. 3.2, those points that intersect with the diagonal line are unmoved after m mappings and are of period m. Their number is 2^m. Among them, there are pseudo period-m points such as $X_n = 0$ which is in fact a fixed point under the map. Apparently, the point of any large period can exist. The number of periodic points is infinite and they can be in one-to-one correspondence with the positive integers, *i.e.*, they are countable. The number of chaotic points is infinite. However, they are uncountable.

Periodic points are dense in $[0, 1]$. They can be anywhere. In Fig. 3.2, for a very small interval $[\varepsilon, 2\varepsilon]$, as long as p is large enough satisfying $2^{-p} < \varepsilon/2$, then in $[\varepsilon, 2\varepsilon]$, there will be a point of period p. In summary, periodic points and chaotic points are mingled together, just like that rational and irrational numbers are in $[0, 1]$. One just cannot *separate* them!

The properties of the 1-dimensional map shown above are universal. They are the common characteristics of a nonlinear system and are very useful for our understanding of nonlinear phenomena in physics.

3.2   Lyapunov exponent in Hamiltonian system

For a conserved system, the volume of its phase space is a constant. This is due to the Hamilton's equations that, if it expands in certain direction, it must contract in the other direction. Hence, chaos is common in Hamiltonian systems. The KAM theorem as described in Chapter 1 states that chaos emerges mainly around the unstable fixed point of pendulum-like *islands*.

One crucial property of a Hamiltonian system is that its Lyapunov exponents are paired. For instance, for a 2-dimensional Hamiltonian system, there are two sets of action-angle variables, i.e., four variables. The energy conservation reduces the degree of freedom to three for which there are three Lyapunov exponents. Since they have to be paired, they can only be of the form: $\lambda_1, 0, -\lambda_1$. Hence, we only need to calculate one exponent. Often, only the maximal Lyapunov exponent needs calculation since it dominates the degree of chaos. Appendix of this chapter shows its calculation.

For the convenience of viewing a trajectory in a multi-dimensional phase space, one may reduce the space dimension by choosing a surface section called **Poincare surface of section** such that a point is marked on it every time the trajectory passes through it from a designated direction, for instance, by specifying $p_i > 0$ or < 0.

3.3   Period 3 route to chaos

Sarkovskii theorem tells that the appearance of periodic trajectories can also imply chaos.

Sarkovskii arranged the positive integers as follows. First the odd numbers except 1, then 2 times, 2^2 times, ..., these numbers. Finally, the powers of 2 in a decreasing order and 1, *i.e.*,

$$3, 5, 7, \ldots, 2 \times 3,\ 2 \times 5,\ 2 \times 7, \ldots, 2^2 \times 3,\ 2^2 \times 5,\ 2^2 \times 7, \ldots,$$

$$2^5,\ 2^4,\ 2^3,\ 2^2,\ 2,\ 1.$$

The sequence is called Sarkovskii sequence.

Sarkovskii theorem tells that in a 1-dimensional system under a continuous map, if there appears a trajectory of period p, then the trajectories of periods that are after p in the Sarkovskii sequence will also appear. Hence, if an odd number period appears, then any periods will appear except those preceding the odd number in the sequence which will not necessarily appear. In particular, the appearance of period-3 trajectory implies the appearance of trajectories of all periods. That is, 'period 3 implies chaos'. (Conceptually, a chaotic trajectory is of period-infinity or just of no period.) In other words, periodic trajectory of period 3 is a route to chaos!

3.4   Resonance overlapping and sine circle mapping

Resonance overlapping may lead to chaos. This can be understood from Fig. 3.3. Resonance overlapping means the adhesion of two pendulum dynamics in the phase space. Therefore, a trajectory can diffuse from one pendulum to another one in a free and chaotic way. This is called the Chirikov diffusion.[1] Shown in Fig. 3.3 is also the Arnold diffusion which appears as the nonlinear coupling is enhanced by the KAM theorem. We have addressed it in Chapter 1. This process has to encounter the hyperbolic point which will slow down the diffusion. Hence, its diffusion rate is slower than that of Chirikov diffusion.

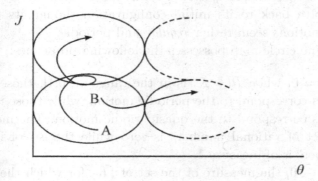

Fig. 3.3   Trajectory A diffuses vertically to the resonance line (Chirikov diffusion). Trajectory B diffuses along the resonance line (Arnold diffusion).

Resonance overlapping can be modelled by the sine circle map. Consider two independent periodic motions with (angular) frequencies Ω_1 and Ω_2. Denote $\Delta\theta_1$ as the evolution of θ_1 of the motion of frequency Ω_1 after a complete cycle of 2π by the motion of frequency Ω_2. $\Delta\theta_1 = 2\pi(\Omega_1/\Omega_2) \equiv W$. We call $R = \Omega_1/\Omega_2$ the rotational number. It shows, as Ω_2 motion evolves one cycle, how many cycles Ω_1 motion will evolve. Consider this process as a mapping and denote the map variable as θ_n on its n-th mapping (which is in fact the value of θ_1 of Ω_1 motion). Then, if there is a nonlinear coupling of sine functional form between these two motions, we have

$$\theta_{n+1} = \theta_n + \Delta\theta_1 = \theta_n + 2\pi(\Omega_1/\Omega_2) + k\sin\theta_n$$

where k is the coupling coefficient. This mapping is called the sine circle map.

Under coupling, R has to be averaged as:

$$\overline{R} = \frac{1}{2\pi}\lim_{m\to\infty}\frac{1}{m}\sum_{n=0}^{m-1}\Delta\theta_n$$

$$= \frac{1}{2\pi}\lim_{m\to\infty}\frac{1}{m}\sum_{n=0}^{m-1}W + k\sin\theta_n.$$

If $\overline{R}$ is rational, then the system motion is periodic; if $\overline{R}$ is irrational, then the motion is quasiperiodic which means that the system will never evolve back to its initial configuration though its two constituent motions seem rather *regular* and periodic.

The sine circle map possesses the following properties:

(1) As $k = 0$, when $\overline{R}(= R)$ is in the range of $[0,1]$, those rational points correspond to the periodic motion, while those irrational points correspond to the quasiperiodic motion. The measure of the set of rational numbers is zero while the set of irrational numbers is of measure 1.
(2) As $k \to 0$, the measure of the set of $\{\overline{R}\}$ for which the motions are quasiperiodic, is still 1. This means that the coupling will not destroy all the quasiperiodic motions immediately.

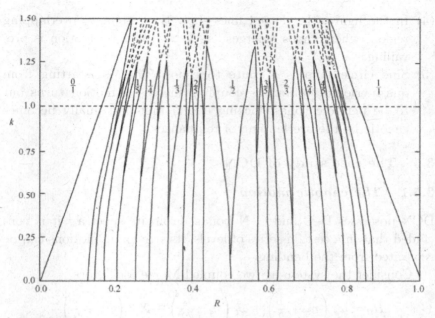

Fig. 3.4 The formation of Arnold tongues under sine circle map. The region in the tongues is the quasiperiodic motion while out of the region is the periodic motion. Dotted region shows the overlap of resonances. Note that a tongue as shown is not a complete tongue, instead, in which there are infinitely many tongues. (adopted from Jensen M H, Bak P, Bohr T. Phys. Rev. A, 1984, 30: 1960)

(3) As k increases, quasiperiodic motions are destroyed gradually and forced to become periodic. As $k = 1$, the measure of the set $\{\overline{R}\}$ for which the motions are quasiperiodic is 0 while that for the periodic motions is 1. Then, except some isolated points whose $\overline{R}$'s are irrational, there is a huge number of points whose $\overline{R}$'s are rational. In other words, then most motions are periodic. This shows the phenomenon of *phase-locking*. Figure 3.4 shows this process diagrammatically. Shown in the figure are the so-called *Arnold tongues* for which $\overline{R}$ is irrational. (Note that a tongue as shown is not a complete tongue, instead, in which there are infinitely many tongues.) The region out of the tongues corresponds to the periodic motion with rational $\overline{R}$. It is the resonance region.

(4) In the figure, the dotted lines show the resonance overlapping region, where chaos emerges. As $k > 1$, chaotic motion is prevailing.

(5) Sine circle map is a route to chaos. That is: starting from quasiperiodic motion, as coupling increases, motion turns out to be periodic (phase-locking effect) first and finally becomes chaotic by the overlapping of resonances.

3.5 The case study of DCN

3.5.1 *The chaotic motion*

DCN possesses D–C and C–N bonds. From its spectra,[2] it is concluded that in a certain series of levels, its stretching motions can be separated from the bending.

Consider the system as two coupled Morse oscillators:

$$H_0 = \omega_s \left(n_s + \frac{1}{2} \right) + \omega_t \left(n_t + \frac{1}{2} \right) + X_{ss} \left(n_s + \frac{1}{2} \right)^2$$
$$+ X_{tt} \left(n_t + \frac{1}{2} \right)^2 + X_{st} \left(n_s + \frac{1}{2} \right) \left(n_t + \frac{1}{2} \right).$$

Here, ω stands for the harmonic frequency, X the anharmonic coefficient. s and t are for the D–C and C–N stretchings, respectively.

The coupling terms can be determined by the fit of the eigenenergies of the algebraic Hamiltonian to the experimental data. The results are that there are two main resonances, 1:1 and 2:3, i.e.,

$$H_{st} = K_{st}(a_s^+ a_t + h.c.)$$
$$H_K = K(a_s^+ a_s^+ a_t a_t a_t + h.c.).$$

The coefficients are listed in Table 3.1. Note that ω_s/ω_t is 1.38 which is between 1 and 1.5. Hence, there are 1:1 and 2:3 resonances.

Table 3.1 The coefficients (in cm^{-1}) of the algebraic Hamiltonian of DCN.

ω_s	ω_t	X_{ss}	X_{tt}	X_{st}	K_{st}	K
2681.4	1948.9	−21.2	−0.3	−34.3	−88.7	13.8

For H_{st}, $P_1 = n_s + n_t$ is conserved. For H_K, $P_2 = n_s/2 + n_t/3$ is conserved. For $H = H_0 + H_{st} + H_K$, no quantities are conserved except energy.

In the coset space (q_s, p_s, q_t, p_t) we have:

$$n_s = (q_s^2 + p_s^2)/2, \quad n_t = (q_t^2 + p_t^2)/2$$

$$H_{st} = K_{st}(q_s q_t + p_s p_t)$$

$$H_K = K\left[(q_s^2 - p_s^2)\left(\sqrt{2}q_t^3 - \frac{3n_t}{\sqrt{2}}q_t\right) - 2q_s p_s\left(\sqrt{2}p_t^3 - \frac{3n_t}{\sqrt{2}}p_t\right)\right].$$

Hamilton's equations are:

$$\partial H/\partial p_\alpha = dq_\alpha/dt, \quad \partial H/\partial q_\alpha = -dp_\alpha/dt \quad (\alpha = \text{s}, \text{t}).$$

For each level, we can obtain its solution space from $H(q_\alpha, p_\alpha) = E$ with E the state energy. Then, we randomly choose 200 initial points to calculate the trajectories. Since the system energy is conserved and positive and negative Lyapunov exponents are paired, we need only to calculate the maximal one, λ.

For a trajectory, if $\lambda > 0$, then we call it chaotic. If $\lambda = 0$, then it is regular. For each level, we calculate 200 λ's and take their average, $\langle\lambda\rangle$. The magnitude of $\langle\lambda\rangle$ will show the degree of chaos of an eigenstate. The result is shown in Fig. 3.5. The levels below 12,000 cm^{-1} are regular. Higher levels up to 18,000 cm^{-1} show significant increase of chaoticity. For levels in this energy range, chaotic and regular trajectories coexist in the phase space. For levels above 18,000 cm^{-1}, all the trajectories are chaotic.

For levels up to between 20,000 cm^{-1} and 30,000 cm^{-1}, the chaoticity reaches its maximum. For levels higher than this energy, the chaoticity decreases instead. This demonstrates the phenomenon of resonance overlapping. The argument is as follows. For levels below 20,000 cm^{-1} and above 30,000 cm^{-1}, 1:1 and 2:3 resonances are dominant, respectively. This is evidenced by the fact that for levels in these two regions, as they are classified by P_1 and P_2 (though they are not exact constants of motion), respectively, Dixon dips are distinct. While for levels in between 20,000 cm^{-1} and 30,000 cm^{-1}, no apparent Dixon dips are observed. The implication is that, in this energy region, both 1:1 and 2:3 resonances are operative, i.e., they

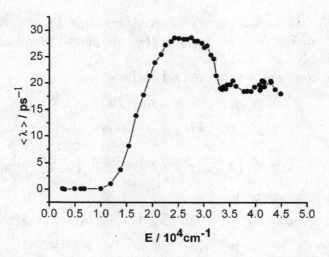

Fig. 3.5   The averaged Lyapunov exponents $\langle \lambda \rangle$ for DCN stretching motions.

are overlapping. This resonance overlapping leads to the increase of chaoticity and $\langle \lambda \rangle$ reaches its maximum as shown by calculation. We will analyze this behavior more in Chapter 6.

3.5.2   *Periodic trajectories*

We note that nonlinearity induces not only chaos but also periodicity. The induced periodicity is different from the periodicity by linearity and possesses peculiar properties. In a nonlinear system, chaotic and periodic trajectories are closely related to each other.

DCN is such a case. At high energy, its vibrational phase space is full of chaotic *sea* in which numerous periodic trajectories are embedded. Noteworthy is the appearance of period-3 trajectories. By Sarkovskii's inference, the appearance of a period-3 trajectory implies the existence of chaos.

In the followings, we will address the period-1, 5, 3, 7, 8, 9, 12, 15 and 18 trajectories. For convenience, the periods are defined by the observation on Poincare surface of section $(q_s, p_s)(q_t = 0, \ p_t < 0)$. As observed from another surface of section such as $(q_t, p_t)(q_s = 0, \ p_s < 0)$, the periods may vary. But this is not essential physically.

In the followings, the level numberings are based on the calculation with $n_s + n_t \leq 10$.

Period-1, 5 trajectories

For the 11th level, there are two period-1 trajectories. Each trace on the (q_s, p_s) and (q_t, p_t) surfaces spans a deviation less than 5×10^{-3} in 10 ps ($1\,\text{ps} = 10^{-12}$ sec). We will stick to this tolerance in the following discussion. Figs. 3.6 (a) and (b) show their phases. ($\phi_\alpha = \tan^{-1}(-p_\alpha/q_\alpha)$, $\alpha = s, t$), respectively. In Fig. 3.6(a), ϕ_s and ϕ_t are in phase and the mode is symmetric. The mode shown in Fig. 3.6(b) is antisymmetric since ϕ_s and ϕ_t are out of phase. Their periods are 17.2 fs and 10 fs, respectively.

The bond displacement Δr corresponding to (q, p) is obtained through the following formula (See Section 1.2) if we consider both D–C and C–N as Morse oscillators:

$$\Delta r = a^{-1} \ln\{[1 - (1 - \lambda^2)^{1/2} \cos\phi]/\lambda^2\}.$$

Here, $a = (-2X\mu)^{1/2}$, μ the reduced mass, $\lambda = 1 - (2n + 1)/k$, $k = \omega/(-X)$, $n = (q^2 + p^2)/2$. Figures 3.6 (c) and (d) are the evolving Δr_s and Δr_t. The mode character, symmetric or antisymmetric, is vivid. Figure 3.7 shows ν_1 and ν_2 normal modes under the harmonic approximation. ν_1 is symmetric while ν_2 is antisymmetric. The amplitude ratios A_s/A_t for these period-1 trajectories are 0.63 and 6.37 which are different from those of ν_1 and ν_2: 0.21 and 1.44. The action (proportional to the square of amplitude) ratios of D–C and C–N are 0.04:1 and 2.07:1 for ν_1 and ν_2. For the symmetric period-1 trajectory, this ratio is 0.40:1 while for the antisymmetric period-1 trajectory, it is 40.58:1. The high localization of the antisymmetric period-1 trajectory is evident. Figures 3.6(e) and (f) show the Fourier transforms of their q_s and q_t. For the symmetric and antisymmetric period-1 trajectories, there are peaks at 1914 cm^{-1} and 2528 cm^{-1}, respectively. These are different from ν_1, ν_2, which are at 1925 cm^{-1} and 2630 cm^{-1}.

In the 23rd level, we observe a symmetric period-5 trajectory. Figure 3.8(a) is the relation of its ϕ_s, ϕ_t. The period is 85.0 fs. On (q_s, p_s) and (q_t, p_t) surfaces, there are five points. The Fourier transforms of q_s and q_t show only one peak at 1914 cm^{-1} which is broader

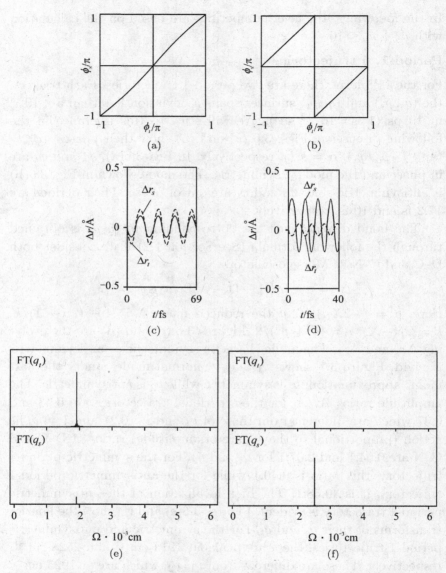

Fig. 3.6   The phases, bond displacements and Fourier transforms of two period-1 trajectories in level 11.

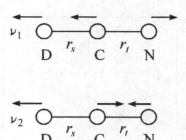

Fig. 3.7 ν_1 and ν_2 normal modes based on the harmonic approximation.

than those of period-1 trajectories. This is not unexpected since their periods are different.

Period-3 trajectories

Figure 3.9(a) shows that on the (q_s, p_s) surface there are two period-3 trajectories in level 22. One is in the *inner* region and the other in the *outer* region. (A similar situation appears in period-1, 7, 9 trajectories.) Figure 3.9(b) shows what is observed on the (q_t, p_t) surface. (Since there are four points on this surface, we may consider it of period 4.) Figures 3.8(b) and (c) show their phase relations. In a period, Δr_s has 4 periods, Δr_t has 3 periods. The periods of these two period-3 trajectories are 56 fs and 52 fs, respectively. In a period, there is time interval, t_s, in which Δr_s and Δr_t are in phase and the interval, t_a, in which Δr_s and Δr_t are out of phase. For the inner period-3 trajectory of the 11th level, t_s/t_a is 1.69. This value increases to 1.96 for the 30th level. This shows that period-3 trajectories are composed of ν_1 and ν_2 with a ratio of 2:1. (Note that Δr_s, Δr_t of ν_1 and ν_2 are in and out of phase, respectively.) The action ratio on D–C and C–N is 2.15:3 $(2 \times 0.04 + 1 \times 2.07 : 2 \times 1 + 1 \times 1)$. For higher levels, the component of ν_1 in period-3 trajectory increases. For the outer period-3 trajectory, t_s/t_a drops from 1.08 of level 11 to 0.96 of level 22. Hence, this period-3 trajectory is composed of ν_1 and ν_2 with a ratio of 1:1. Indeed, the action ratio, n_s/n_t, on D–C and C–N is close to 1 $(1 \times 0.04 + 1 \times 2.07 : 1 \times 1 + 1 \times 1)$. The Fourier spectrum also shows that inner period-3 trajectory has two peaks at 1914 cm^{-1}

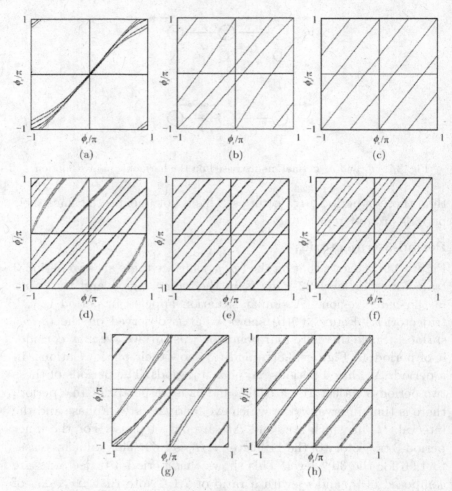

Fig. 3.8 The phases of various periodic trajectories (a) period-5, level 23; (b) period-3, level 22; (c) period-3, level 22; (d) period-7, level 28; (e) period-7, level 22; (f) period-8, level 15; (g) period-9, level 28; (h) period-9, level 20.

and 2553 cm^{-1}. For the outer period-3 trajectory, the peaks are at 1878 cm^{-1} (Ω_t) and 2503 cm^{-1} (Ω_s) accompanied by other smaller peaks at 3191 cm^{-1}($2\Omega_s - \Omega_t$) and 1253 cm^{-1}($2\Omega_t - \Omega_s$). These peaks are different from ν_1 and ν_2. The appearance of period-3 trajectories as viewed on (q_s, p_s) implies that D–C motion is chaotic.

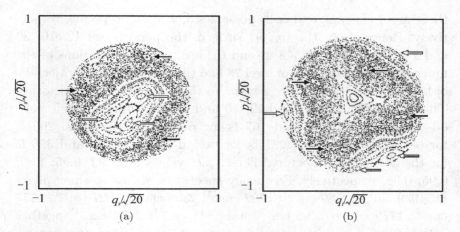

Fig. 3.9   Period-3 trajectories of level 22 on the (q_s, p_s) and (q_t, p_t) surfaces. Arrows show their locations.

Period-7, 8 trajectories

The period-7 trajectory as shown on the (q_s, p_s) surface is of period 9 if viewed on the (q_t, p_t) surface. The phase relations of the inner period-7 trajectory of level 28 and the outer period-7 trajectory of level 22 are shown in Figs. 3.8(d),(e). Though their phase relations are different, their periods are close to 119 fs and t_s/t_a are 1.53 and 1.08, respectively. The Fourier spectra of q_s and q_t show peaks at 1878 cm^{-1}, 2418 cm^{-1} and 1927 cm^{-1}, 2479 cm^{-1} together with more complicated minor components.

The period-8 trajectory on (q_s, p_s) becomes period-11 on (q_t, p_t). For instance, the period-8 trajectory of level 15 possesses a period of 148 fs. The phase relation is shown in Fig. 3.8(f). t_s/t_a is 0.83 Frequency peaks are at 1792 cm^{-1} and 2467 cm^{-1}.

These period-9, 11 trajectories on the (q_t, p_t) surface, according to Sarkovskii theorem, also imply chaos of the C–N stretching motion.

Period-9, 12, 15, 18 trajectories

The trajectories of periods that are multiples of 3 (like 9, 12, 15 and 18) possess characteristics similar to the period-3 trajectory. On the (q_s, p_s) surface, the corresponding 9, 12, 15 and 18 points aggregate into three clusters of 3, 4, 5 and 6 points, respectively. On (q_t, p_t), there are 4 clusters, and each possesses 3, 4, 5 and 6 points, respectively. Hence, from the (q_t, p_t) surface, the periods are 12, 16, 20 and 24. Shown in Figs. 3.8(g) and (h) are the phase relations of the inner period-9 trajectory of level 28 and the outer period-9 trajectory of level 20. Indeed, they are *similar* to those of period-3 trajectories. This is the same for period-12, 15 and 18 trajectories. Their periods are longer: 154 fs and 167 fs for period-9 trajectories, 206 fs for period-12 trajectory, 281 fs for period-15 trajectory and 326 fs for the period-18 trajectory. Their t_s/t_a values are:1.92, 0.95; 1.89; 0.95; 0.98, respectively. Frequency spectra of q_s and q_t show peaks at: 1939 cm^{-1}, 2589 cm^{-1}; 1792 cm^{-1}, 2393 cm^{-1}; 1927 cm^{-1}, 2577 cm^{-1}; 1779 cm^{-1}, 2368 cm^{-1} and 1841 cm^{-1}, 2454 cm^{-1} together with minor complicated combinations. As the period of a trajectory increases, the trajectory also approaches chaos and its frequency spectrum becomes more complicated as shown in Fig. 3.10.

Just like ν_1, in-phase of Δr_s and Δr_t implies that more the C–N stretching is involved and the larger t_s/t_a is. Conversely, like ν_2, out-of-phase of Δr_s and Δr_t implies that more the D–C stretching is involved and the smaller t_s/t_a is. For the period-3 trajectory of level 15, $t_s/t_a = 1$. Therefore, if a trajectory is in the outer region of the period-3 trajectory, then its $t_s/t_a < 1$. Otherwise, it is in the inner region and its $t_s/t_a > 1$. Chaotic trajectories are in the outer region of the period-3 trajectory with t_s/t_a in the range of 0.7–0.95. In other words, chaotic trajectories involve more D–C motion.

However, close to the center of the (q_t, p_t) surface as shown in Fig. 3.9(b), there is evidence that energy of high excitation is localized on the quasi-periodic mode of the D–C stretching. This mode is on the outer rim of the (q_s, p_s) surface. Figure 3.11 shows the relative distribution of actions on D–C (n_s) and C–N (n_t) of this mode. Indeed, action is concentrated on D–C and highly localized.

In summary, regardless of the period, periodic trajectories are composed of ν_1 and ν_2 in a ratio of 2:1 or 1:1 and the action ratios on D–C and C–N are 2:3 and 1:1. These are just in conformity with

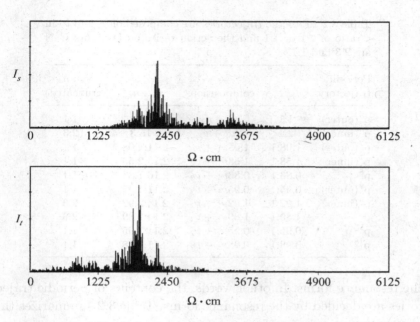

Fig. 3.10   Frequency spectra of a chaotic trajectory.

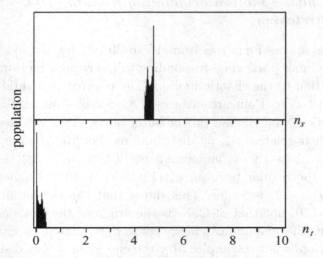

Fig. 3.11   The relative distribution of actions on D–C (n_s) and C–N (n_t), showing that action is concentrated on D–C and highly localized.

Table 3.2   Periodic trajectories are composed of ν_1 and ν_2 in a ratio of 2:1 or 1:1 and the action ratios on D–C and C–N are 2:3 and 1:1.

Periodic trajectory	t_s/t_a	compositions	$n_s : n_t$	$n_s : n_t$ (truncated)
p^3(outer)	1:1	$\nu_1 + \nu_2$	2.11:2	1:1
p^3(inner)	2:1	$2\nu_1 + \nu_2$	2.15:3	2:3
p^7(outer)	1.08:1	$1.08\nu_1 + \nu_2$	2.11:2.08	1:1
p^7(inner)	1.53:1	$1.53\nu_1 + \nu_2$	2.13:2.53	~1:1
p^8	0.83:1	$0.83\nu_1 + \nu_2$	2.10:1.83	~1:1
p^9(outer)	0.95:1	$0.95\nu_1 + \nu_2$	2.11:1.95	1:1
p^9(inner)	1.92:1	$1.92\nu_1 + \nu_2$	2.14:2.92	2:3
p^{12}	1.89:1	$1.89\nu_1 + \nu_2$	2.14:2.89	~2:3
p^{15}	0.95:1	$0.95\nu_1 + \nu_2$	2.11:1.95	1:1
p^{18}	0.98:1	$0.98\nu_1 + \nu_2$	2.11:1.98	1:1

the resonance forms. In other words, the contents of periodic trajectories are decided by the resonance forms. Table 3.2 summarizes this observation.

3.5.3   *Chaotic motion originating from the D–C stretching*

The phase spaces for levels from 16 to 30 can be divided into two parts: the inner portion corresponding to the regular motion and the outer portion to the chaotic motion. The relative area ratio is fixed. Figure 3.12 shows Poincare surfaces of section of some regular (in the inner portion) and chaotic (in the outer portion) trajectories of level 19 and their relative n_s, n_t distributions. For the chaotic portion, n_s $(4.4 > n_s > 1.9)$ is larger and n_t $(3.4 > n_t > 0.2)$ is smaller while for the regular portion, n_s $(1.2 > n_s > 0)$ is smaller and n_t $(5.8 > n_t > 4.3)$ is larger. This shows that the excitation of D–C (action >1.9), not that of C–N, is the origin of the chaotic motion. Of course, the amount of energy excited is related to chaos. But the situation is not so simple. How the energy or action distribution among the bond modes can also be an important factor for chaotic motion to appear.

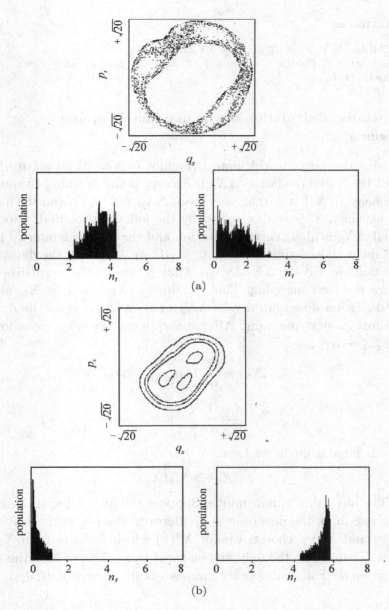

Fig. 3.12 Poincare surfaces of section $(q_s, p_s)(p_t < 0, q_t = 0)$ of some chaotic trajectories (a) and regular trajectories (b) of level 19 of DCN and their relative n_s, n_t distributions.

References

1. Chirikov B V. *Phys. Rep.*, 1979, 52: 263.
2. Baggott J E, Caldow G L, Mills I M. *J. Chem. Soc. Faraday Trans.*, 1988, 2(84): 1407.

Appendix: Calculation of the maximal Lyapunov exponent

In using computer to calculate Lyapunov exponent, no matter how small the initial deviation ΔX_0 is chosen, if the maximal Lyapunov exponent $\lambda_1(X_0) > 0$, then very soon, ΔX_n will be beyond the memory capacity. This can be solved by the following method. Suppose initial ΔX_0 is along certain direction and the mapping interval is τ. At time τ (initial time is 0), $\Delta X_1 = \alpha_1$. At X_1, along the deviation direction, we divide ΔX_1 by α_1. That is, we do the normalization before the next mapping. Then, we have $\Delta X_2 = \alpha_2$. At X_2, along the deviation direction, divide ΔX_2 by α_2 and use it as the deviation for the next mapping. After n mappings, the total deviation is $\alpha_n \alpha_{n-1} \cdots \alpha_1$, i.e.

$$\lambda_{n1}(X_0) = \frac{1}{n\tau} \ln \alpha_n \alpha_{n-1} \cdots \alpha_1$$

$$= \frac{1}{n\tau} \sum_{r=1}^{n} \ln \alpha_r.$$

As n is large enough, we have:

$$\lambda_1(X_0) \cong \lambda_{n1}(X_0).$$

The algorithm is not unique. Suppose the initial deviation is d_0, after one map, the deviation is d_1. Between the two evolved points, $X_1(\tau)$ and $X'(\tau)$, choose a point $X''(\tau)$ which deviates from $X_1(\tau)$ by d_0. Then we do the mapping on these two points to have the new deviation d_2. Follow this procedure, we obtain a series of $d_1, d_2, \ldots d_n$ and

$$\lambda_{n1}(X_0) = \frac{1}{n\tau} \sum_{r=1}^{n} \ln(d_r/d_0).$$

This is a generic formula. As $d_0 = 1$, we retrieve the previous formula.

Sometimes, we find that λ_{n1} varies with n in the form of $\alpha n^{-\beta}$ $(\alpha, \beta > 0)$ or that $\ln \lambda_{nj}$ and $\ln n$ follow a linear relation of $\ln \alpha - \beta \ln n$. Then as $n \to \infty$, λ_{n1} will approach zero. That is, the Lyapunov exponent is zero.

The calculation of the rest Lyapunov exponents is more cumbersome. However, often, only this largest Lyapunov exponent is what we need since it determines mainly the degree of chaos.

Chapter 4

C–H Bending Motion of Acetylene

4.1 Introduction

Acetylene has two carbon atoms and two hydrogen atoms. Its C–H bond is single and the C–C bond is triple. The H atom will migrate to another C atom to form vinylidene if its bending motion is energetically high enough. This process can be modeled as the 1,2-shift in organic chemical reactions. The way to reach high bending motion by light absorption is impractical due to relaxation. The SEP (stimulated emission pumping) and dispersed fluorescence technique are first used to excite the molecule to its higher electronic states (X state), then as it relaxes (either via stimulated emission or fluorescence) to the electronic ground states, its vibrational or rotational states can be in high excitation.

Acetylene has five modes: symmetric C–H stretching (labeled as 1), C–C stretching (2), antisymmetric C–H stretching (3), *trans* C–H bending (4) and *cis* bending (5). SEP and dispersed fluorescence experiments show that the highly excited vibrational states obtained are via the couplings of Darling–Dennison (DD) and vibrational l doubling due to the C–H bending motion (l is used to denote both the angular momentum and its quantum number). In the early stage of the formation of these highly excited states, quantum numbers $N_{str} = n_1 + n_2 + n_3$, $N_{res} = 5n_1 + 3n_2 + 5n_3 + n_4 + n_5$ and

$l = l_4 + l_5$ are conserved under these resonances. (Note that n, N, l are notations for quantum numbers).

Among these highly excited states, those with $N_{str} = 0$ which are pure C–H bending are of particular interest. This is related to the 1,2 migration. For this case, $N_{res} = n_4 + n_5 \equiv N_b$ is conserved. Currently, the C–H bending state up to 15000 cm^{-1} has been assigned for which N_b is close to 22. This energy is about that required for H migration which is about 48 kcal/mol. The equivalent n_4 or n_5 is 23–28.

4.2   Empirical C–H bending Hamiltonian

The empirical Hamiltonian H_{eff} can be written with its coefficients determined from the fit to the experimental data. They are listed in Table 4.1.[1]

H_{eff} has the explicit form:

(1) The diagonal elements:

$$\langle n_4 l_4, n_5 l_5 | H_{\text{eff}} | n_4 l_4, n_5 l_5 \rangle$$
$$= \omega_4 n_4 + \omega_5 n_5 + x_{44} n_4^2 + x_{55} n_5^2 + x_{45} n_4 n_5 + g_{44} l_4^2 + g_{45} l_4 l_5 + g_{55} l_5^2.$$

Here, ω shows the mode frequency in cm^{-1}. x, g are nonlinear coefficients.

(2) The off-diagonal terms:

(a) DD-I:

$$\langle n_4 l_4, n_5 l_5 | H_{\text{eff}} | (n_4 - 2) l_4, (n_5 + 2) l_5 \rangle$$
$$= s_{45}/4 [(n_4^2 - l_4^2)(n_5 + l_5 + 2)(n_5 - l_5 + 2)]^{1/2}$$

Table 4.1   The coefficients of the empirical Hamiltonian in cm^{-1}.

ω_4	ω_5	x_{44}	x_{45}	x_{55}	g_{44}	g_{45}	g_{55}
608.656	729.137	3.483	−2.256	−2.389	0.676	6.671	3.535

y_{444}	y_{445}	y_{455}	y_{555}	r_{45}^0	r_{445}	r_{545}	s_{45}
−0.0306	0.0241	0.0072	0.00954	−6.193	0.0303	0.0109	−8.572

(b) DD-II

$$\langle n_4 l_4, n_5 l_5 | H_{\text{eff}} | (n_4 - 2)(l_4 \mp 2), (n_5 + 2)(l_5 \pm 2) \rangle$$
$$= (r_{45} + 2g_{45})/16[(n_4 \pm l_4)(n_4 \pm l_4 - 2)(n_5 \pm l_5 + 2)$$
$$\times (n_5 \pm l_5 + 4)]^{1/2}$$

(c) vibrational l doubling:

$$\langle n_4 l_4, n_5 l_5 | H_{\text{eff}} | n_4(l_4 \pm 2), n_5(l_5 \mp 2) \rangle$$
$$= r_{45}/4[(n_4 \mp l_4)(n_4 \pm l_4 + 2)(n_5 \pm l_5)(n_5 \mp l_5 + 2)]^{1/2}.$$

Here, $r_{45} = r_{45}^0 + r_{445}(n_4 - 1) + r_{545}(n_5 - 1)$.

4.3 Second quantization representation of H_{eff}

Rewrite n_4, l_4, n_5, l_5 as:

$$n_4 = n_{4+} + n_{4-}, \quad n_5 = n_{5+} + n_{5-}$$
$$l_4 = n_{4+} - n_{4-}, \quad l_5 = n_{5+} - n_{5-}.$$

The subscripts "+", "−" can be understood as the right and left rotational motions. The relation between n and l is:

$$l_i = n_i, n_i - 2, \ldots, -n_i, \quad i = 4, 5.$$

The conserved l, N_b are:

$$l = l_4 + l_5 = n_{4+} + n_{5+} - n_{4-} - n_{5-}$$
$$N_b = n_4 + n_5 = n_{4+} + n_{5+} + n_{4-} + n_{5-}.$$

(Note that for H_{eff} in the last section, l and N_b are indeed conserved.)
For the "+" and "−" motions, we have the constants:

$$P_1 \equiv (N_b + l)/2 = n_{4+} + n_{5+}$$
$$P_2 \equiv (N_b - l)/2 = n_{4-} + n_{5-}.$$

This shows that the quantum numbers for "+" and "−" motions are conserved, respectively.

In the new variables $(n_{4+}, n_{4-}, n_{5+}, n_{5-})$, the diagonal elements of H_{eff} are trivial. However, the off-diagonal elements are very compact. For instance, DD-I changes to:

$$s_{45}[n_{4+}n_{4-}(n_{5+} + 1)(n_{5-} + 1)]^{1/2}$$

which is:

$$s_{45}(a_{4+}^+ a_{4-}^+ a_{5+} a_{5-} + h.c.)$$

in terms of the second quantized operators. Similarly, for DD-II and vibrational l doubling, we have:

$$(r_{45} + 2g_{45})/4(a_{4+}^+ a_{4+}^+ a_{5+} a_{5+} + h.c.)$$
$$(r_{45} + 2g_{45})/4(a_{4-}^+ a_{4-}^+ a_{5-} a_{5-} + h.c.)$$

and

$$r_{45}(a_{4-}^+ a_{4+} a_{5+}^+ a_{5-} + h.c.).$$

4.4 su(2) ⊗ su(2) represented C–H bending motion

For "+" (the same for "−" bending) bending, we can define:

$$J_{x+} = (a_{4+}^+ a_{5+} + a_{5+}^+ a_{4+})/2$$
$$J_{y+} = -i(a_{4+}^+ a_{5+} - a_{5+}^+ a_{4+})/2$$
$$J_{z+} = (n_{4+} - n_{5+})/2.$$

It is easy to show that $\{J_{x+}, J_{y+}, J_{z+}\}$ satisfy su(2) algebra (The algebra that the angular momenta satisfy). By these operators, DD-I and vibrational l doubling can be combined to form a very simple term:

$$4s_{45}J_{x+}J_{x-}.$$

Here, we approximate $r_{45} \approx s_{45}$ since their values, $-6.19\,\text{cm}^{-1}$ and $-8.57\,\text{cm}^{-1}$, are indeed very close.

Meanwhile, DD-II can be written as:

$$(r_{45} + 2g_{45})[(J_{x+}^2 + J_{x-}^2) - (n_4 + n_5)/4 - (n_4 n_5 + l_4 l_5)/4].$$

We note that the last two terms can be grouped into the diagonal elements.

For modeling, we can simplify the diagonal elements (though this is not necessary) as:

(1) $x_{45}n_4n_5 + g_{45}l_4l_5 - (r_{45} + 2g_{45})(n_4n_5 + l_4l_5)/4$

simplified to

$$x_1(n_{4+}n_{5+} + n_{4-}n_{5-})$$

(2) $x_{44}n_4^2 + x_{55}n_5^2 + g_{44}l_4^2 + g_{55}l_5^2$

simplified to

$$x_2(n_{4+}^2 + n_{4-}^2 + n_{5+}^2 + n_{5-}^2)$$

x_1 and x_2 are parameters.

(3) $\omega_4 n_4 + \omega_5 n_5 - (r_{45} + 2g_{45})(n_4 + n_5)/4$

simplified to

$$\omega_4(n_{4+} + n_{4-}) + \omega_5(n_{5+} + n_{5-})$$

since $(r_{45} + 2g_{45}) \ll \omega_4, \omega_5$, so the corresponding term can be omitted.

In summary, we have a very compact algebraic Hamiltonian $H_{\text{algebraic}}$

$$\omega_4(n_{4+} + n_{4-}) + \omega_5(n_{5+} + n_{5-}) + x_1(n_{4+}n_{5+} + n_{4-}n_{5-})$$
$$+ x_2(n_{4+}^2 + n_{4-}^2 + n_{5+}^2 + n_{5-}^2) + 4s_{45}J_{x+}J_{x-}$$
$$+ (r_{45} + 2g_{45})(J_{x+}^2 + J_{x-}^2).$$

This Hamiltonian is built up by two coupled "+" and "−" su(2) algebras.

$J_{x+}J_{x-}$ and $J_{x+}^2 + J_{x-}^2$ are of second order. From a symmetry consideration, there is one first order term $\lambda(J_{x+} + J_{x-})$. As the quantum numbers are small, this term is important. At high excitation, it is not as prominent as the second order terms.

It is interesting to note that "+" and "−" motions are coupled via the nonlinear term $J_{x+}J_{x-}$. The term $J_{x+}^2 + J_{x-}^2$ (or DD-II coupling) will not couple "+" and "−" motions, neither the first order term $J_{x+} + J_{x-}$.

4.5   Coset representation

In the two "+", "−" coset representations, we have coordinates (q_+, p_+, q_-, p_-) and

$$n_{4+} = 2(q_+^2 + p_+^2), \quad n_{5+} = P_1 - n_{4+}$$

$$n_{4-} = 2(q_-^2 + p_-^2), \quad n_{5-} = P_2 - n_{4-}.$$

The diagonal terms of H_{eff} can be easily expressed in terms of (q_+, p_+, q_-, p_-). The off-diagonal terms can be derived from the algebraic expressions as:

DD-I: $4s_{45}(n_{5+}n_{5-})^{1/2}(q_+q_- - p_+p_-)$

DD-II: $(r_{45} + 2g_{45})[n_{5+}(q_+^2 - p_+^2) + n_{5-}(q_-^2 - p_-^2)]$

vibrational l coupling: $4r_{45}(n_{5+}n_{5-})^{1/2}(q_+q_- + p_+p_-).$

By these representations, we have the (q_+, p_+, q_-, p_-) represented H_{eff}. The Hamilton's equations of motion are:

$$\partial H_{\text{eff}}/\partial q_\alpha = -dp_\alpha/dt$$

$$\partial H_{\text{eff}}/\partial p_\alpha = dq_\alpha/dt \quad (\alpha = +, -).$$

The dynamics then is fully described by the trajectories in the two coupled coset spaces for which (q_+, p_+, q_-, p_-) are the coordinates.

Analogously, we may have $H_{\text{algebraic}}(q_+, p_+, q_-, p_-)$. For a given energy (which can be eigenenergy or not) E, we can obtain the solution space (q_+, p_+, q_-, p_-) from $E = H_{\text{algebraic}}(q_+, p_+, q_-, p_-)$ (or $H_{\text{eff}}(q_+, p_+, q_-, p_-)$). The solution space is the phase space.

4.6   Modes of C–H bending motion

Previously, the dynamics of the C–H bending motion of acetylene is realized by two coupled su(2) algebras, $\{J_{x+}, J_{y+}, J_{z+}\} \otimes \{J_{x-}, J_{y-}, J_{z-}\}$. H_{eff}(or $H_{\text{algebraic}}$) adopts the so-called normal mode picture in which 4 and 5 denote the *trans* and *cis* normal modes, respectively. For the su(2) algebra, we can rotate $\pi/2$ along the J_y axis to form the new $\{J_x', J_y', J_z'\}$ such that $J_x' = -J_z, J_y' = J_y$, $J_z' = J_x$. In the new system, $4', 5'$ (those with super prime notation) denote the bending motions of the two C–H bonds, respectively.

This is the so-called local mode picture for which we have the (q'_+, p'_+, q'_-, p'_-) space. The advantage of the local mode picture is its intuitiveness. We can first obtain the solution space (q_+, p_+, q_-, p_-), then shift via transformation to the (q'_+, p'_+, q'_-, p'_-) space to obtain quantities like $n'_4, n'_5, \ldots$, etc.. The transformation has been shown in Section 2.4.

Current experiments are limited to the cases with $l = 0, 2$. Our calculation shows that the results for the cases with zero and nonzero l are not much different. In the following discussion, we will limit ourselves to the case where $l = 0$. In such case, $P_1 = P_2$ and the system dynamics is symmetric with respect to '+' and '−' motions. Therefore, only the discussion for the '+' case is necessary. We will stress the relations between $\Delta n_+ (\equiv (n_{4+} - n_{5+})/P_1)$ and $\varphi_+ (\equiv \tan^{-1}(-p_+/q_+)$ is the angle), $\Delta n_- (\equiv (n_{4-} - n_{5-})/P_2)$ and $\varphi_-, \Delta n_-$ and $\Delta n_+, \varphi_-$ and φ_+ (The same is true for the local mode picture). One point that needs clarification is that different initial points will lead to different trajectories. Our discussion will be from the viewpoint of the *global* properties. Though trajectories are different, they can show common *global* properties. Therefore, we will only show representative trajectories for discussion. In particular, we will compare the behaviors of the lower and higher levels for various N_b.

(a) $N_b = 6, l = 0$

For $N_b = 6, l = 0$, we have 16 states. They are labeled with L1,L2,...,L16 from the lowest one.

Figure 4.1(a) shows the property of L2. Since $\Delta n_+ \approx P_1$, most action (energy) is stored in the '4' mode, i.e., the *trans* bending. That φ'_+ is around $\pm\pi$ implies the phase difference between the two C–H bending motions is π. Hence, it is antisymmetric. Figure 4.1(b) shows that L15 is a *cis* bending. The two C–H bendings are in phase.

(b) $N_b = 14, l = 0$

For L3 (Fig. 4.1(c)), since Δn_+ is close to 1, the mode is close to a *trans* mode. Meanwhile, φ'_+ is close to $\pm\pi$. So, these two observations are consistent. However, $\Delta n'_+$ is close to ± 1 and the traces form two discrete regions. This shows that action is concentrated on either C–H bending and cannot flow freely between the two C–H bendings. Note

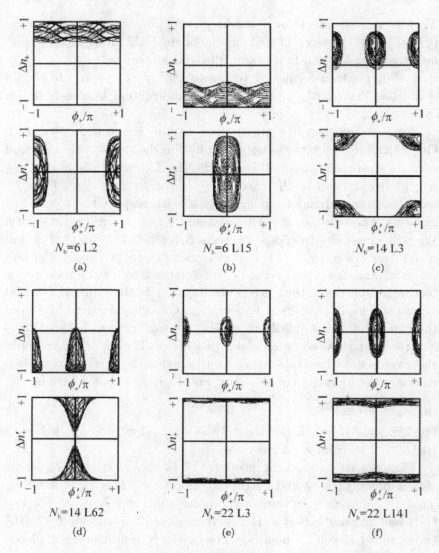

Fig. 4.1   The trajectories in the phase space for different states.

that though this mode is *trans*, it is different from L2 of $N_b = 6$. φ_+ of L2 is extensive between $-\pi$ and π while here φ_+ is limited to 0 and $\pm\pi$. For the high level L62 (Fig. 4.1(d)), the mode is *cis* and possesses localized character. Contrary to L15 of $N_b = 6, \varphi_+$ is limited to 0 and $\pm\pi$.

(c) $N_b = 22, l = 0$

For the low levels like L3, or high levels like L141 (Figs. 4.1(e), (f)), φ'_+ is extensive and $\Delta n'_+$ is nearly 1, -1. This shows that the two C–H bendings are independent with no phase correlation and form a local mode with action concentrated only on either bending. For higher levels like L25, action can flow between the two C–H bendings with φ'_+ being $\pi/2$. As viewed along the C–C bond, when the angle between the two C–H bendings is $\pi/2$, the action transfer is most probable.

In the above discussion, the dynamics of the C–H bending is analyzed by the trajectories. In the following analysis, we will discuss the dynamics from a global viewpoint on the phase space.

(a) Low levels of $N_b = 6, 12, 14, 22$

We will analyze the mode characters by viewing the relations among: $\Delta n_+, \Delta n_- (\Delta n'_+, \Delta n'_-); \varphi_-, \varphi_+ (\varphi'_-, \varphi'_+)$. Figure 4.2(a) shows the result of $N_b = 6$, L $= 2$. Since $\Delta n_+ = (n_{4+} - n_{5+})$ and $\Delta n_- = (n_{4-} - n_{5-})$ are close to the maximal value $N_b/2$, i.e., both n_{5+} and n_{5-} are close to zero, hence, $n_4 = n_{4+} + n_{4-} = N_b$. The mode is *trans* and normal. This also shows up in that the difference between φ'_-, φ'_+ is π. The relation between the corresponding φ_-, φ_+ is not definite. Neither is the relation between $\Delta n'_-$ and $\Delta n'_+$. This shows that action can freely flow between the two C–H bendings. This is the character of a normal mode. This is consistent with the previous analysis.

The case for $N_b = 12$, L3 is shown in Fig. 4.2(b). The mode is *trans*. But, n'_4 or n'_5 is close to the maximum. The action is concentrated on either C–H bending and its transfer between the two C–H bendings is limited. Since φ'_+ and φ'_- are around π, the mode is *trans*. As $N_b = 14$, the mode character of L3 remains the same as shown in Fig. 4.2(c). Now, both φ'_+ and φ'_- show dispersive behavior. The phase angle between the two C–H bendings is no longer fixed at π. This is the character of a local mode.

As $N_b = 22$, the result for L3 is shown in Fig. 4.2(d). n_4 and n_5 are $N_b/2$. Either n'_4 or n'_5 is N_b. There is no fixed phase relation between φ'_- and φ'_+. These indicate that action is localized on either

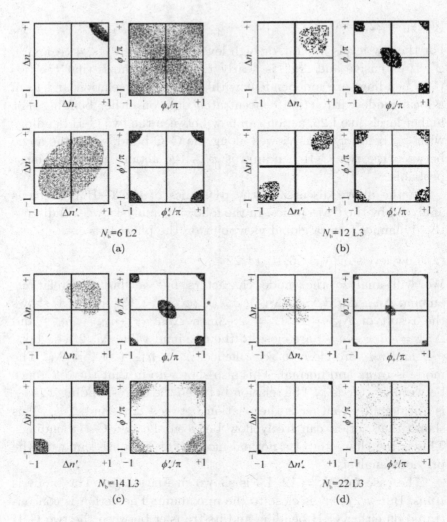

Fig. 4.2 The phase space structures for different states.

C–H. Besides, $n'_{4+} = n'_{4-}$ (or $n'_{5+} = n'_{5-}$) shows that the bending motion of C–H is in a plane. This *one hydrogen* bending motion can be visualized as shown in Fig. 4.3.

(b) High levels of $N_b = 6, 12, 14, 22$

As $N_b = 6$, the result of L15 is shown in Fig. 4.4(a). From the relation of Δn_- and Δn_+, we know that n_5 is close to N_b and n_4 is very small.

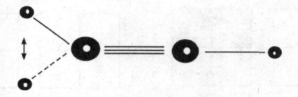

Fig. 4.3   One hydrogen bending motion.

This is a *cis* normal mode. That $\varphi'_+ \approx \varphi'_- \approx 0$ and no fixed relation between φ_- and φ_+ also supports this assertion. The relation of $\Delta n'_+$ and $\Delta n'_-$ also indicates that action can flow between the two C–H bendings.

As $N_b = 12$, the result of L48 is shown in Fig. 4.4(b). n_5 is close to N_b and n_4 is almost 0. This is a *cis* mode. However, a difference with the common normal mode is that the phase between φ_- and φ_+ is fixed around $\varphi_+ = \pi, \varphi_- = 0$ or $\varphi_+ = 0, \varphi_- = \pi$. Besides, n'_{4+} and n'_{5-} are close to $N_b/2$ and correspondingly, n'_{4-} and n'_{5+} are very small; or that both n'_{4-} and n'_{5+} are close to $N_b/2$, while n'_{4+} and n'_{5-} are very small. This shows that the two C–H bendings are in opposite rotation with roughly equal action. The energy transfer between the two C–H is not so free since the relation diagram between $\Delta n'_+$ and $\Delta n'_-$ is bisected into two regions. (This can also be seen from the relation of $\Delta n'_+$ and φ'_+ or $\Delta n'_-$ and φ'_- that around $\varphi'_+ = \varphi'_- = 0$, the diagram (not shown) is separated into two regions) Since $\varphi'_+ = \varphi'_- = 0$, as one C–H is in '+' rotation, though the action of another C–H that is in '+' rotation is very small, they are still in phase. This phenomenon is also true for '−' rotation.

As $N_b = 14$, the result of L62 is shown in Fig. 4.4(c). The two C–H bendings are in opposite rotation. However, the distribution of φ'_+, φ'_- is becoming dispersive. The two C–H bendings are losing their phase relation in '+' or '−' motion. For L141, $N_b = 22$, Fig. 4.4(d) shows that the phase relation no longer exists. Motion is such that the two C–H are in counter rotation, showing evident local character. This particular mode is depicted in Fig. 4.5.

From the aspect of trajectory, there is no apparent difference for L3 and L141 of $N_b = 22$ (See Figs. 4.1(e), (f)). However, from the

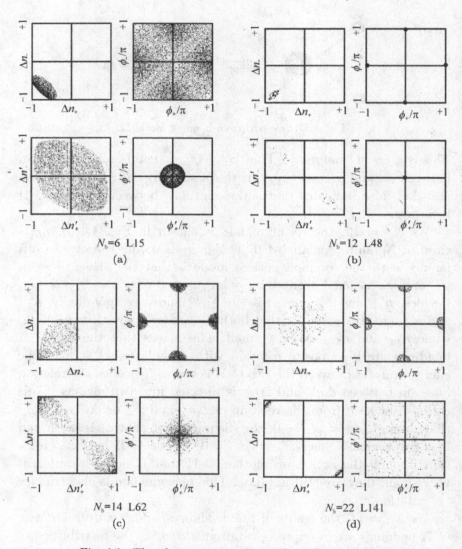

Fig. 4.4   The phase space structures for different states.

global aspect of phase space, they are distinctly different. (Compare the relation of $\Delta n'_+, \Delta n'_-$ in Figs. 4.2(d), 4.4(d)) The difference lies in that for L3, $n'_4 = N_b$ while $n'_5 = 0$ (or $n'_4 = 0, n'_5 = N_b$). For L141, $n'_{4+} = n'_{5-} = N_b/2, n'_{4-} = n'_{5+} = 0$ (or $n'_{4-} = n'_{5+} = N_b/2, n'_{4+} = n'_{5-} = 0$). L3 is one hydrogen bending while L141 is that the two C–H

Fig. 4.5   Counter rotation of two C–H bending motions.

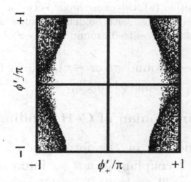

Fig. 4.6   The phase relation for the case with nonzero vibrational angular momentum.

bendings are in counter rotation. Apparently, the global approach is superior to that based on the trajectory analysis. This is expected since trajectories are initial condition dependent and the approach based on them is hardly fully exploited as dynamics is concerned.

We note that the modes as shown in Figs. 4.3 and 4.5 have also been reported[2,3] by a quantal algorithm although without those details about the dynamical configuration and inter-mode action transfer as are shown here.

The vibrational angular momentum l caused by the two C–H bendings is $l'_+ - l'_-$ or $(n'_{4+} + n'_{5+}) - (n'_{4-} + n'_{5-})$. It is the difference between the '+' and '−' rotations. Figure 4.6 shows the case of L4 with $N_b = 6, l = 2$. The '+' motion possesses larger rotational momentum (or angular momentum) since for the '−' motion, the phase angle between the two C–H bendings can range between $-\pi$ and π, i.e., with no definite phase angle, while for the '+' motion, the

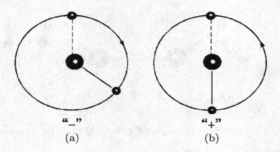

Fig. 4.7   For the '−' motion (a), the phase angle between the two C–H bendings can range between $-\pi$ and π, i.e., with no definite phase angle, while for the '+' motion (b), the phase angle is centred around π (or $-\pi$).

phase angle is centred around π (or $-\pi$), i.e., in *trans* configuration with nonzero l. This is depicted in Fig. 4.7.

4.7   Reduced Hamiltonian of C–H bending motion

As mentioned previously, in the formation of the C–H bending motion, vibrational l coupling is not so important as DD-I, DD-II. In general, l is small (only 0, 2) and can be neglected. Therefore, we consider only the DD coupling which can be simplified as $K_{\mathrm{DD}}(a_4^{+2}a_5^2 + h.c.)$. 4 and 5 stand for the *trans* and *cis* modes. In addition, we consider other modes as a bath for modes 4 and 5. Though modes 4 and 5 are not in resonance with the bath since their frequencies are not compatible, there can be anharmonic coupling among them. Hence, the *reduced* Hamiltonian is:

$$H_{reduced} = H_{bath} + H_{\text{int}} + H_b$$

with

$$H_{bath} = \sum_{i=1}^{3} \omega_i(n_i + 1/2) + \sum X_{ii}(n_i + 1/2)^2$$

$$+ \sum_{i<j} X_{ij}(n_i + 1/2)(n_j + 1/2)$$

$$H_{\text{int}} = \sum_{i=1}^{3} X_{i4}(n_i + 1/2)(n_4 + 1/2) + \sum X_{i5}(n_i + 1/2)(n_5 + 1/2)$$

$$H_b = \omega_4(n_4 + 1/2) + \omega_5(n_5 + 1/2) + X_{44}(n_4 + 1/2)^2$$
$$+ X_{55}(n_5 + 1/2)^2 + X_{45}(n_4 + 1/2)(n_5 + 1/2)$$
$$+ K_{DD}(a_4^{+2}a_5 + h.c.).$$

The coefficients have been determined in the literature by the fit to the experimental level energies.[4,5] They are listed in Table 4.2.

By the coset representation, we have $J_+ = a_4^+ a_5$, $J_- = a_5^+ a_4$, $J_z = n_4 - n_5$ and:

$$n_4 = 2(q^2 + p^2), \quad n_4^2 = 2(q^2 + p^2)\{[J - (q^2 + p^2)]/J + 2(q^2 + p^2)\}$$
$$n_5 = 2[J - (q^2 + p^2)],$$
$$n_5^2 = 2J\{2J + (q^2 + p^2)/[J - (q^2 + p^2)]\}\{[J - (q^2 + p^2)]/J\}^2$$
$$n_4 n_5 = 2(2J - 1)[J - (q^2 + p^2)](q^2 + p^2)/J$$
$$J_+^2 + J_-^2 = 4(2 - J^{-1})[J - (q^2 + p^2)](q^2 + p^2).$$

Table 4.2   The coefficients of the reduced Hamiltonian (cm^{-1}).

ω_1	3398.74
ω_2	1981.71
ω_3	3316.09
ω_4	609.01
ω_5	729.17
X_{11}	−26.57
X_{22}	−7.39
X_{33}	−27.41
X_{44}	−3.08
X_{55}	−2.34
X_{12}	−12.62
X_{13}	−105.09
X_{14}	−15.58
X_{15}	−10.85
X_{23}	−6.10
X_{24}	−12.48
X_{25}	−1.57
X_{34}	−6.96
X_{35}	−8.69
X_{45}	−2.41
K_{DD}	−2.75

For this reduced Hamiltonian, n_1, n_2, n_3 are given integers and n_4, n_5 are treated as continuous variables. The reduced Hamiltonian is in the normal mode picture. Phase angle $\varphi = \tan^{-1}(-p/q)$ is the phase difference between the *trans* and *cis* modes. The transformation from the normal mode picture to the local mode picture (q', p') is straightforward. (See Section 2.4).

For an eigenenergy E_i, we can obtain $q = q(p)$ from the equation:

$$E_i = H_{\text{bath}}(n_1, n_2, n_3) + H_b(q, p) + H_{\text{int}}(n_1, n_2, n_3, q, p).$$

From the solution, the dynamical properties can be easily obtained. The mode characters are summarized below:

(a) Local mode

In the local mode picture, if φ' is extensive from 0 to 2π, then it is a local mode. In the local mode picture, the mode possesses two separate and equivalent trajectories, one with $n_4' > n_5'$, while the other with $n_4' < n_5'$. They are analogous to the clockwise and counterclockwise rotations of a pendulum. In the normal mode picture, the phase φ of the local mode is limited to 0 and π. If in a motion cycle, $n_4 > n_5$, then it possesses more of the *trans* character. Otherwise, if $n_4 < n_5$, then the mode has more of the *cis* character. We label them by L_t and L_c, respectively.

(b) Normal mode

For the normal mode, in the local mode picture, φ' is limited to a specific range and the coupling is strong. While in the normal mode picture, φ is extensive in the whole range of $[0, 2\pi]$. If φ' is centred around 0, then the two C–H bendings are in phase. This mode is called symmetric and denoted by N_s. Otherwise, if φ' is centred around $\pi(-\pi)$, then the mode is antisymmetric and labeled by N_a. In the normal mode picture, N_s and N_a possess *cis* $(n_5 > n_4)$ and *trans* $(n_4 > n_5)$ characters, respectively.

(c) Precessional mode

The precessional mode character has φ centred at $\pi/2(-\pi/2)$ while φ' is centred in between 0 and $\pi(-\pi)$. If $n_4 > n_5$, the mode is labeled by P_t. If $n_4 < n_5$, then it is labeled by P_c. φ' of P_c is centred in

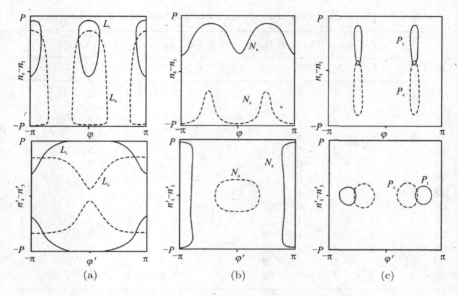

Fig. 4.8 The local (a), normal (b) and precessional (c) modes.

between 0 and $\pi/2(-\pi/2)$. φ' of P_t is centred in between $\pi/2$ and $\pi(-\pi/2, -\pi)$.

These mode characters are depicted in Fig. 4.8.

For the reduced Hamiltonian of the C–H bending system, n_1, n_2, n_3 are given and $P \equiv n_4 + n_5$ is a conserved quantity ($P = 2J$). For a given P, there are $P/2 + 1$ levels. In general, for a two-mode system with 1:1 and 2:2 couplings, low levels are of local character while high levels are of normal character. This is true for the C–H bending motion of acetylene though with variations. Table 4.3 shows the distribution of mode characters for different P under $(n_1, n_2, n_3) = (0, 0, 0)$. In the Table, the level energy is higher across from L_t, L_c, N_a, N_s, P_t to P_c. For $P < 12$, all levels are normal. The lower levels are N_a and the higher ones are N_s. The increase of P only results in the birth of more levels of N_s character. As P is between 14 and 18, levels of N_a character disappear and the number of N_s levels remains unchanged. Meanwhile, the number of L_t levels increases. As P reaches 22, both L_t and N_s levels disappear, accompanied by the birth of L_c, N_a and P_c, P_t levels. As P is even larger,

Table 4.3   The distribution of mode characters for different P and n_1, n_2, n_3.

| | $(n_1,n_2,n_3)=(0,0,0)$ | | | | | | $(n_1,n_2,n_3)=(5,0,0)$ | | | | | |
P	L_t	L_c	N_a	N_s	P_t	P_c	L_t	L_c	N_a	N_s	P_t	P_c
4		1	2					1	2			
6		2	2					2	2			
8		2	3					2	3			
10		2	4					2	4			
12		2	5					2	5			
14	2		6					2	6			
16	3		6				2		7			
18	4		6				3		7			
20	5		4		2		4		7			
22		7	1		4		5		5			2
24		7	3	3			7		3			3
26		7	4	3			8		2			4
28		7	6	2			9	2		4		
30		7	7	2			8	5		3		

| | $(n_1,n_2,n_3)=(0,5,0)$ | | | | | | $(n_1,n_2,n_3)=(0,0,5)$ | | | | | |
P	L_t	L_c	N_a	N_s	P_t	P_c	L_t	L_c	N_a	N_s	P_t	P_c
4		1	2					1	2			
6		2	2					2	2			
8		2	3					2	3			
10		2	4					2	4			
12		3	4				2		5			
14		3	5				2		6			
16		3	6				3		6			
18		2	8				5		3			2
20	3		8				6		2			3
22	4		8				7	2		3		
24	5		8				6	4		3		
26	6		6		2		6	6		2		
28	7		5		3		6	7		2		
30	9		3		4		6	8		2		

there remain only the L_c, N_a and P_t levels and only the number of N_a levels increases; the rest do not increase evidently.

With C–H stretching excitation, like $(n_1,n_2,n_3)=(5,0,0)$, $(0,0,5)$, the results are shown also in Table 4.3. It seems that the mode distributions for various P follow the same trend. For the C–C

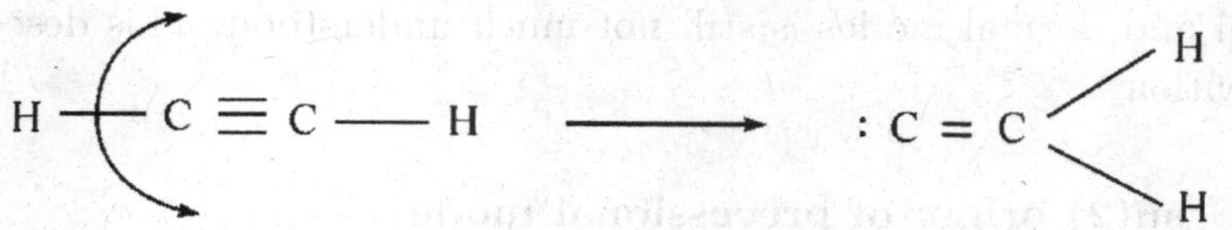

Fig. 4.9 The transition from acetylene to vinylidene.

excitation, like $(n_1, n_2, n_3) = (0, 5, 0)$, the distribution is also shown therein. As $P = 24 - 30$, there are more L_t levels than in the case only with C–H stretching excitation. As P is larger, the number of L_t levels also increases by a large amount.

The transition from acetylene to vinylidene is depicted in Fig. 4.9. The potential barrier by the quantum chemical calculation is about $48 \, \text{kcal/mole}$[6] which is equivalent to about 23–28 quanta on the C–H bending. For such a transition, there are three criteria:

(1) There must be 23–28 quanta on the C–H bending excitation. The modes must be local. Otherwise, the excitation will be dispersed between the two C–H bendings. This will lead to action decrease on each C–H bending and the transition to surmount the barrier will be more difficult.

(2) For the local modes, L_t mode is more appropriate since *trans* configuration is adequate for the migration of an H atom from one carbon atom to the other without steric hindrance. Therefore, the formation of vinylidene is more probable.

(3) From the experimental viewpoint, we will have more choices if there are more L_t levels.

From the above consideration, we expect that the accompanying C–C stretching excitation is helpful to the transition from acetylene to vinylidene.

There are two points that need attention here. One is that as the frequency of one (or several) mode(s) is far from the rest modes, it is appropriate to consider these modes as a bath with which the mode(s) is anharmonically coupled. This concept could be helpful to the understanding of intramolecular vibrational energy redistribution (IVR). The second is that not much is known about the precessional mode. What its role is in IVR as contrasted with the local and normal modes is still not much understood. This deserves attention.

4.8   su(2) origin of precessional mode

Why is there no precessional mode in the eigenstates of H_{eff} or $H_{\text{algebraic}}$ while in the reduced Hamiltonian the precessional mode appears? We will try to understand this from an algebraic viewpoint.

We know that the coupling of two modes α and β can be:

$$a_\alpha^+ a_\beta + a_\alpha^+ a_\alpha^+ a_\beta a_\beta + \cdots + h.c.$$

For this system, we can construct an su(2) algebra $\{J_x, J_y, J_z\}$ as:

$$J_x = (a_\alpha^+ a_\beta + a_\beta^+ a_\alpha)/2,$$
$$J_y = -i(a_\alpha^+ a_\beta - a_\beta^+ a_\alpha)/2,$$
$$J_z = (n_\alpha - n_\beta)/2.$$

In terms of the su(2) operators, the interaction is

$$C_1 J_x + C_2 J_x^2 + \cdots.$$

We can rotate $\pi/2$ along the J_y axis so that $J_z' = J_x$, i.e., the quantization is along the x axis. The transformation between $\{J_x, J_z\}$ and $\{J_z', -J_x'\}$ leads to the local and normal mode pictures.

In principle, we can have $C_3 J_y$ (or J_y^2) interaction for which the phase angles of the trajectories, in the normal and local mode pictures, are centred around $\pi/2$. This is the precessional mode as shown in Fig. 4.8(c). Figure 4.10 shows the normal, local and precessional modes in the normal and local mode pictures.

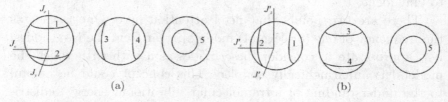

Fig. 4.10   (a) Normal mode picture, (b) Local mode picture. 1,2 represent the normal modes, 3,4 the local modes and 5 the precessional mode.

DD-I, DD-II and vibrational l doubling in H_{eff} can be written in terms of J_x, J_y as:

$$\text{DD-I:}\quad 2s_{45}(J_{x+}J_{x-} - J_{y+}J_{y-})$$

$$\text{DD-II:}\quad 1/2(r_{45} + 2g_{45})(J_{x+}^2 + J_{x-}^2 - J_{y+}^2 - J_{y-}^2)$$

$$\text{vibrational } l \text{ doubling:}\quad 2r_{45}(J_{x+}J_{x-} + J_{y+}J_{y-}).$$

The coefficient of DD-II is smaller than those of DD-I and vibrational l doubling ($g_{45} \sim -r_{45}$, hence $(r_{45} + 2g_{45})/2 \sim -r_{45}/2$). Besides, the high order terms of DD-II cancel each other out. For simplicity, DD-II is neglected. Since $s_{45} \approx r_{45}$, $J_{y+}J_{y-}$ terms in DD-I and vibrational l doubling cancel each other out. Only the $J_{x+}J_{x-}$ term remains in H_{eff}. Hence, there are only normal and local modes and no precessional mode.

To confirm this assertion, let $r_{45} = -s_{45}$, then only the $J_{y+}J_{y-}$ term remains in H_{eff}. Indeed, there is a change from Fig. 4.11(a) with $r_{45} = s_{45}(n_4 + n_5 = 6, l = 0)$ to (b) with $r_{45} = -s_{45}(n_4 + n_5 = 22, l = 0)$. In Fig. 4.11(b), trajectories are only around the J_y axis and the precessional mode appears. (Note that Fig. 4.11 is true for both '+' and '−' motions.)

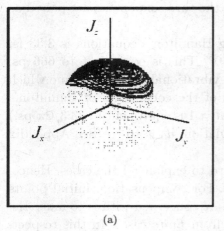 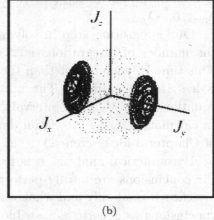

(a)   (b)

Fig. 4.11   (a) The trajectories with $r_{45} = s_{45}(n_4 + n_5 = 6, l = 0)$ and (b) with $r_{45} = -s_{45}(n_4 + n_5 = 22, l = 0)$. In (b), trajectories are only around the J_y axis and there appears the precessional mode.

In summary, we have shown the theoretical background of the precessional mode. That is: in an su(2) algebraic Hamiltonian, J_x, J_z axes are the limit of the local and normal modes while the J_y term leads to the precessional mode. Finally, we note that precessional mode was first mentioned by Noid and Marcus in their semiclassical calculation of the Henon–Heiles model with 1:1 coupling.[7]

4.9 Nonergodicity of C–H bending motion

For each eigenenergy E, by the coset represented $H_{eff}(q_+, p_+, q_-, p_-)$, we can obtain the solution space from $H_{eff}(q_+, p_+, q_-, p_-) = E$. From a point in the solution space, by Hamilton's equations of motion, a trajectory can be followed. The trajectory is three dimensional due to energy conservation. For convenience, the surface of section $(q_+, p_+, q_- = 0, p_- < 0)$ can be chosen for viewing the trajectory, *i.e.*, as $q_- = 0$, and $p_- < 0$, the corresponding (q_+, p_+) values are recorded. (For convenience, q_+, p_+ are normalized by $\sqrt{N_b/2}$ to the range of $[-1, 1]$. We will only consider the case of $l = 0$). The Lyapunov exponent for each trajectory is also calculated. For this system, we only have to calculate the largest Lyapunov exponent, λ_{max} because the three exponents are paired as $\lambda_{max}, 0, -\lambda_{max}$.

Our integration step in solving Hamilton's equations is 3.33 fs, the number of integrations is $2*10^5$. This is equivalent to 666 ps. This time is much longer than the vibrational relaxation time which is less than $1 - 10$ ps. (The unit of the coefficients of Hamilton's equations is cm^{-1}. Time interval $\Delta t = 1$ is equivalent to $33.3/2\pi$ ps.) In calculating λ_{max}, the initial deviation, d_0 is 10^{-6}. (See Appendix of Chapter 3 for reference.)

By numerical analysis, it is hard to exploit all the cases. Hence, the conclusions are hardly perfect. For compensation, initial points are chosen randomly and what we are concerned with is to make the conclusions as generic as possible from *finite* cases. In this respect, numerical analysis is still beneficial.

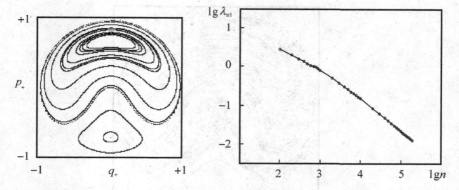

Fig. 4.12 The surface of section of a representative trajectory of L9 $(3997.1\,\text{cm}^{-1})$ and the relation between its $\log \lambda_{n1}$ and $\log n$. Obviously, its form is $\alpha n^{-\beta}$ and $\lambda_{\max} = 0$.

Numerical conclusions show that the C–H bending motion can be:

(1) Regular motion

For $N_b = 6$ (N_b is the total action or quantum number of bending motion), there are 16 levels and these are labeled from the lowest one with L# (# is $1, \ldots, (N_b/2+1)^2$). All these levels possess zero $\lambda_{\max}$, showing that their trajectories are regular. Figure 4.12 shows the surface of section of a representative trajectory of L9 $(3997.1\,\text{cm}^{-1})$ and the relation between its $\log \lambda_{n1}$ and $\log n$. Obviously, its form is $\alpha n^{-\beta}$ and $\lambda_{\max} = 0$.

(2) Irregular motion

For each N_b between 8 and 22, except those higher and lower levels for which trajectories are regular, the phase spaces are full with trajectories with various $\lambda_{\max}$. For example, for $N_b = 10$, L17 $(6602.2\,\text{cm}^{-1})$, $\lambda_{\max}$ ranges from 0.18 ps^{-1} to 0.36 ps^{-1}. Its surface of section is shown in Fig. 4.13 which is full of complex structure.

Figure 4.14 shows the surfaces of section of two trajectories and the relations of their $\log \lambda_{n1}$ and $\log n$ for $N_b = 12$, L21 $(7900.7\,\text{cm}^{-1})$. Their $\lambda_{\max}$ are 0.86 ps^{-1} and 0.32 ps^{-1}, respectively.

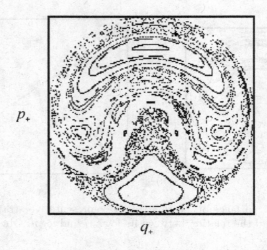

Fig. 4.13   The complex surface of section of $N_b = 10$, L17 ($6602.2\,\mathrm{cm}^{-1}$).

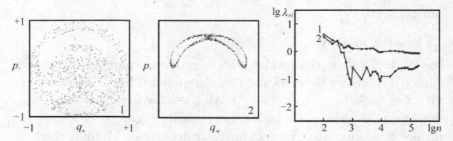

Fig. 4.14   The surfaces of section of two trajectories and the relations of their $\log \lambda_{n1}$ and $\log n$ for $N_b = 12$, L21 ($7900.7\,\mathrm{cm}^{-1}$).

Since there is a distribution of $\lambda_{\max}$, the motion is not ergodic. For each N_b, $\lambda_{\max}$ increases in general from lower to higher levels and then decreases. Also, as N_b is larger, $\lambda_{\max}$ is larger. For a quantitative concept, for a given N_b, we record the maximal $\lambda_{\max}$ of all its levels and denote it as Λ (in ps^{-1}). We have $(\Lambda, N_b) = (0.06, 8)$, $(0.36, 10)$, $(0.93, 12)$, $(1.56, 14)$, $(2.16, 18)$, $(2.43, 22)$. Λ is larger as N_b increases.

Though the general cases are nonergodic, there are cases of ergodicity. For example, for $N_b = 14$, L8 ($9070.9\,\mathrm{cm}^{-1}$), all the trajectories in the phase space share the same $\lambda_{\max}$. Figure 4.15 shows the traces of three trajectories and their plots of $\log \lambda_{n1}$ against $\log n$. Indeed, they share the same $\lambda_{\max}$.

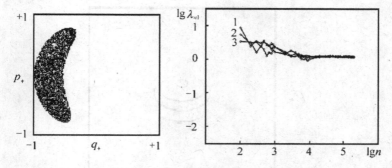

Fig. 4.15 The traces and the plots of $\log \lambda_{n1}$ against $\log n$ for three different trajectories that share the same $\lambda_{\max}$.

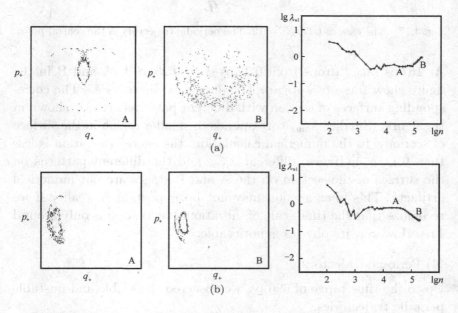

Fig. 4.16 Two cases of the transition between different degrees of chaos.

(3) Transition between different degrees of chaos

An interesting case is the transition between different degrees of chaos for a trajectory as time elapses. Figure 4.16 shows two cases:

Figure 4.16(a) is the case with $N_b = 14$, L46 (9435.7 cm^{-1}). At 300 ps. $\lambda_{\max}$ jumps from 0.43 ps^{-1} to 0.93 ps^{-1}. The degree of chaos enhances. Figure 4.16(b) is the case with $N_b = 18$, L6 (11576.7 cm^{-1}).

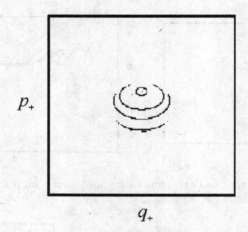

p_+

q_+

Fig. 4.17   The case of L47, $N_b = 12$. The periodic trajectory is the central point.

At 167 ps, $\lambda_{\max}$ drops from $0.72\,\mathrm{ps}^{-1}$ to $0.25\,\mathrm{ps}^{-1}$. A and B in the figure show these two stages. The degree of chaos varies. The corresponding surfaces of section with different patterns are also shown in the figure. (Smaller $\lambda_{\max}$ corresponds to smaller region in the surface of section). In the numerical calculation, the energy deviation is less than $0.1\,\mathrm{cm}^{-1}$. Hence, different $\lambda_{\max}$ and the different patterns on the surface of the section on the A and B stages are not numerical artifacts. This phenomenon may not be important in reality if we recognize that the time scale of vibrational relaxation is only around 1 ps. However, its physics is noticeable.

(4) Periodic trajectories

Up to the time range of 666 ps, we observed the stable and unstable periodic trajectories.

(a) Stable periodic trajectories in $N_b = 12$, L47 ($8373.5\,\mathrm{cm}^{-1}$) and $N_b = 6$, L9 ($3997.1\,\mathrm{cm}^{-1}$). Viewed from the surface of section, the periodic trajectories are immersed in the quasi-periodic trajectories. Figure 4.17 shows the case of L47. The periodic trajectory is the central point. Its $\lambda_{\max}$ is 0.

Figure 4.18 shows the relation of $n'_{4-} - n'_{5-} \equiv \Delta n'_-$ and $n'_{4+} - n'_{5+} \equiv \Delta n'_+$ for L47. (4, 5 denote the two C–H bendings. + and − denote the right and left rotations. The superscript ' ' denotes the

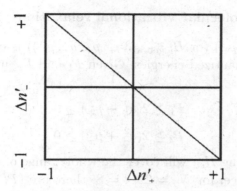

Fig. 4.18 The relation of $n'_{4-} - n'_{5-} \equiv \Delta n'_-$ and $n'_{4+} - n'_{5+} \equiv \Delta n'_+$ for L47.

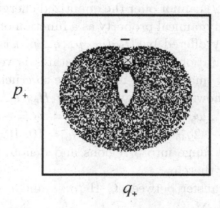

Fig. 4.19 The surface of section and the unstable periodic trajectory marked with $\times$, which is among the chaotic sea for $N_b = 14$, L46 (9435.7 cm^{-1}).

local mode picture.) Since the slope is -1, $n'_{4+} - n'_{5+} = -(n'_{4-} - n'_{5-})$ or $n'_4 = n'_5$, i.e., the actions on the two C–H bendings are the same. For the case of L9 (not shown), the slope is 1 and $l'_4 = l'_5$, i.e., the vibrational angular momenta of the two C–H bendings are always the same.

(b) Unstable periodic trajectory in $N_b = 14$, L46 (9435.7 cm^{-1}). Figure 4.19 shows the surface of section and the unstable periodic trajectory marked with $\times$, which is among the chaotic sea. Note its $\lambda_{max} = 1.18$ ps^{-1} and $n'_4 = n'_5$. That is, the actions on the two C–H bendings are the same.

4.10   Intramolecular vibrational relaxation

$H_{\text{eff}}(q_+, p_+, q_-, p_-)$ (or $H_{\text{algebraic}}(q_+, p_+, q_-, p_-)$) is classical. It cannot offer the quantized energies. Given P_1 and P_2, under the conditions that

$$P_1 \geq 2(q_+^2 + p_+^2) \geq 0$$
$$P_2 \geq 2(q_-^2 + p_-^2) \geq 0$$

the corresponding H_{eff} will cover the whole range of quantized energies. For instance, for $N_b = 22$, $l = 8$, there are $(P_1 + 1)(P_2 + 1) = [(N_b + l)/2 + 1] \cdot [(N_b - l)/2 + 1] = 128$ levels. The energy range of these 128 levels corresponds to that covered by H_{eff}. Though $H_{\text{eff}}(q_+, p_+, q_-, p_-)$ cannot offer the quantized energies, if we are only interested in the dynamical property as a function of energy, then the continuous energy offered by $H_{\text{eff}}(q_+, p_+, q_-, p_-)$ is just enough. In fact, at high excitation, the density of states is very high and the concept of energy quantization may not be so crucial.

Figure 4.20 shows the energy range by $H_{\text{eff}}(q_+, p_+, q_-, p_-)$ with $\omega_4 = 608.66\,\text{cm}^{-1}$, $\omega_5 = 729.14\,\text{cm}^{-1}$, $x_1 = x_2 = -3.0\,\text{cm}^{-1}$, $4s_{45} = -34.28\,\text{cm}^{-1}$, $r_{45} + 2g_{45} = 7.14\,\text{cm}^{-1}$, $N_b = 10, 16, 22$ and $l = 0, 8$. We partition the range into 5 regions and denote them by various notions as shown therein.

The action transfer between C–H *trans* and *cis* bending modes in a time interval Δt is:

$$\Delta n_{45} = 4 \left(\sum_{\alpha} q_\alpha dq_\alpha / dt + p_\alpha dp_\alpha / dt \right) \Delta t$$

$$= 4 \left(\sum q_\alpha \Delta H_{algebraic} / \Delta p_\alpha - p_\alpha \Delta H_{algebraic} / \Delta q_\alpha \right) \Delta t$$

$$\equiv d_{45} \Delta t \quad (\alpha = `+`, `-`).$$

For a point (q_+, p_+, q_-, p_-) in the phase space, we have the corresponding d_{45}. As d_{45} is positive, we suppose the action is transferred from the *cis* bending to the *trans* bending and the process is reversed when d_{45} is negative. (It does not matter if this definition is in an alternative way) Now we consider those (q_+, p_+, q_-, p_-) for which d_{45} is negative. For each Δt, we calculate $d_{45}\Delta t$. As $|d_{45}\Delta t / N_b| < 0.01$,

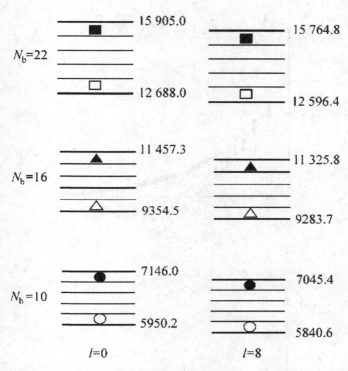

Fig. 4.20 The energy range by the algebraic $H_{\text{eff}}(q_+, p_+, q_-, p_-)$ and their partitions.

we consider that action is still in the *trans* mode and set $P(\Delta t) = 1$, otherwise (that is $|d_{45}\Delta t/N_b| > 0.01$), we consider that action is no longer in the *trans* mode and set $P(\Delta t) = 0$. $P(\Delta t)$ is the survival probability of the *trans* mode as a function of Δt. Of course, we have to average the probability over all points in the phase space. From now on, $P(\Delta t)$ means the averaged survival probability. It is an important dynamical parameter showing the action transfer rate between the *trans* and *cis* modes. Figure 4.21 shows $P(\Delta t)$ for each energy range denoted in Fig. 4.20.

From the figure, we note that except the very high levels of $N_b = 22, l = 0$, $P(\Delta t)$ is not so much dependent on how high a level is and the value of N_b. (Careful study reveals that as l is larger, $P(\Delta t)$ tends to be smaller. But for $N_b = 10$, it tends to be larger.) The relaxation time is roughly between 0.45 and 0.6 ps. An exception is that for the

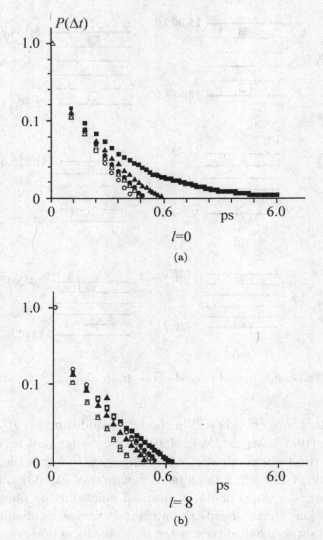

Fig. 4.21 $P(\Delta t)$ for each energy range denoted in Fig. 12.1 (a) $l = 0$ (b) $l = 8$.

highest level of $N_b = 22, l = 0$, the relaxation time can extend up to 10 fold, or 6 ps. This implies that faster relaxation is not a necessary consequence of high excitation. The core cause is the nonlinearity of inter-mode couplings.

For a system with slower relaxation, its spectrum will be *simpler*, such that the spectral peaks will be narrower, discrete and easier to be recognized. Indeed, in the dispersed fluorescence experiment, as N_b is close to 22, the spectrum due to the relaxation (transition) of the *trans* mode to the *cis* mode, i.e., the fractionation of the *trans* mode on the *cis* mode (this is the overlap of the wave function of the *trans* mode on the *cis* mode) is much simpler than in the other cases. Quantum calculation also indicates that as N_b is large, the energy transfer between the *trans* and *cis* modes shows quasiperiodic pattern.[1,2,8] All these are consistent with our classical results by the coset algorithm.

The conclusions are: (1) For high excitation, the classical algorithm can be suitable in some cases. There is no dogma that a strict quantal algorithm has to be followed. (2) The coset algorithm is simple and it grasps the core of the complicated issue.

References

1. Jacobson M P, O'Brien J P, Silbey R J, Field R W. *J. Chem. Phys.*, 1998, 109: 121.
2. Jacobson M P, Silbey R J, Field R W. *J. Chem. Phys.*, 1999, 110: 845.
3. Jacobson M P, Jung C, Taylor H S, Field R W. *J. Chem. Phys.*, 1999, 111: 600.
4. Jonas D M, Solina S A B, Rajaram B, Silbey R J, Field R W, Yamanouchi K, Tsuchiya S. *J. Chem. Phys.*, 1993, 99: 7350.
5. Solina S A B, O'Brien J P, Field R W, Polik W F. *J. Phys. Chem.*, 1996, 100: 7797.
6. Chen W, Yu C. *Chem. Phys. Lett.*, 1997, 277: 245.
7. Noid D W, Marcus R A. *J. Chem. Phys.*, 1977, 67: 559.
8. Jacobson M P, O'Brien J P, Field R W. *J. Chem. Phys.*, 1998, 109: 3831.

Chapter 5

Assignments and Classification of Vibrational Manifolds

5.1 Formaldehyde case

One characteristic of molecular highly excited vibration is its immense number of levels and high density of states. Due to inter-mode interactions, many quantum numbers or constants of motion are destroyed. However then, there are still some good quantum numbers left. These remaining quantum numbers are called the *polyad* numbers. We note that polyad numbers are only approximate *locally*. That is, they are conserved only in certain dynamical processes. This is because as time elapses, the dynamics will become very complicated and the resonance forms may vary so that the polyad number as an operator may no longer commute with the full Hamiltonian.

For instance, formaldehyde has six vibrational modes among which the out-of-plane vibration is quite decoupled from the remaining five in-plane modes. We may concentrate on the vibrational system composed of these five in-plane modes. They are two equivalent C–H stretchings, denoted by the subscripts 1, 5, the C–O stretching by the subscript 2, scissor bending of CH_2 by the subscript 3 and the C–O wagging by the subscript 6. The couplings among these five modes can be represented by the algebraic Hamiltonian:

$$K_{15}(a_1^+ a_5 + h.c.) + K_{23}(a_2^+ a_3 + h.c.) + K_{36}(a_3^+ a_6 + h.c.)$$
$$+ K_{26}(a_2^+ a_6 + h.c.)$$
$$+ K_{122}(a_1^+ a_2 a_2 + a_5^+ a_2 a_2 + h.c.)$$
$$+ K_{133}(a_1^+ a_3 a_3 + a_5^+ a_3 a_3 + h.c.)$$
$$+ K_{166}(a_1^+ a_6 a_6 + a_5^+ a_6 a_6 + h.c.).$$

Here, *h.c.* stands for hermitian conjugate. K_{ij}'s are the coupling strengths.

Under these resonances, actions n_1, n_2, n_3, n_5, n_6 are no longer good quantum numbers. However, it is easy to check that

$$P = 2n_1 + n_2 + n_3 + 2n_5 + n_6$$

is a conserved quantity.

These five modes can be considered as coupled Morse oscillators with resonances. The Morse part Hamiltonian can be written as:

$$
\begin{aligned}
H_0 = {} & \omega_1(n_1 + n_5 + 1) + \omega_2(n_2 + 1/2) + \omega_3(n_3 + 1/2) + \omega_6(n_6 + 1/2) \\
& + X_{11}[(n_1 + 1/2)^2 + (n_5 + 1/2)^2] + X_{22}(n_2 + 1/2)^2 \\
& + X_{33}(n_3 + 1/2)^2 + X_{66}(n_6 + 1/2)^2 \\
& + X_{12}(n_1 + n_5 + 1)(n_2 + 1/2) + X_{13}(n_1 + n_5 + 1)(n_3 + 1/2) \\
& + X_{15}(n_1 + 1/2)(n_5 + 1/2) + X_{16}(n_1 + n_5 + 1)(n_6 + 1/2) \\
& + X_{23}(n_2 + 1/2)(n_3 + 1/2) + X_{26}(n_2 + 1/2)(n_6 + 1/2) \\
& + X_{36}(n_3 + 1/2)(n_6 + 1/2).
\end{aligned}
$$

The coefficients of the Hamiltonian can be determined by fitting the eigenenergies of this algebraic Hamiltonian to the experimental values. There were such reports as tabulated in Table 5.1.[1,2]

For a given P, there are many associated states. All these states share a common quantum number P. Like for $P = 8$, there are 176 states. They can be calculated by this algebraic Hamiltonian, first by constructing the Hamiltonian matrix of 176×176 dimensions through choosing 176 bases of $|n_1, n_2, n_3, n_5, n_6\rangle$ corresponding to $P = 8$. Then, by diagonalizing this Hamiltonian matrix, the

Table 5.1   The coefficients (cm^{-1}) of algebraic Hamiltonian for formaldehyde.

ω_1	2885.66
ω_2	1735.45
ω_3	1539.46
ω_6	1264.90
K_{15}	-36.05
K_{23}	-62.22
K_{26}	-0.14
K_{36}	86.47
K_{122}	9.23
K_{133}	2.28
K_{166}	3.61
X_{11}	-56.94
X_{12}	0.03
X_{13}	-22.32
X_{15}	-0.51
X_{16}	119.36
X_{22}	-11.89
X_{23}	-1.66
X_{26}	37.50
X_{33}	-19.82
X_{36}	27.76
X_{66}	-18.86

176 eigenenergies and states can be elucidated. Each eigenstate $|\varphi\rangle$ is a linear combination of the basis states:

$$|\varphi\rangle = \sum C |n_1, n_2, n_3, n_5, n_6\rangle$$

$|n_1, n_2, n_3, n_5, n_6\rangle$ are called the zeroth-order eigenstates. This is because they are the eigenstates when all K_{ij} are zero (i.e., without any resonances). In this situation, n_1, n_2, n_3, n_5, n_6 are good quantum numbers.

5.2   Diabatic correlation, formal quantum number and level reconstruction

Shown in the right most column in Fig. 5.1(a) are the 176 levels of formaldehyde with $P = 8$. No order of the levels is evident. In the following, we will try to seek the level order. For this, the retrieval

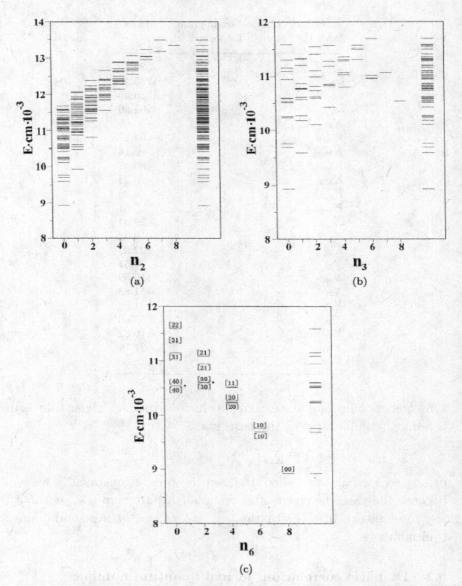

Fig. 5.1   The 176 levels of formaldehyde with $P = 8$ and their orders by various actions. • denotes that the energy spacing is very small.

of the approximate quantum numbers or constants of motion is the crucial step. These quantum numbers are *approximate* since they are not the real or exact ones due to the inter-mode couplings. In other words, we will see that though these quantum numbers are not exact or destroyed by the couplings, they are still useful since by them, levels can be reconstructed to show regularity. For their retrieval, we first introduce the concept of diabatic correlation.

In quantum mechanics, it is well known that as the interaction between two levels varies, the levels will avoid crossing each other. Along the noncrossing route is the so-called adiabatic correlation, while along the route which leads to crossing is the diabatic correlation.

For the seven resonances of formaldehyde, we can switch them off one by one. Correspondingly, there is avoidance of crossing. At each of these points we stick to the diabatic correlation. This will finally lead to the correlation of each level of the full Hamiltonian with a zeroth-order level of H_0. The eigenfunctions of H_0 are $|n_1, n_2, n_3, n_5, n_6\rangle$. Therefore, formally, we can label the levels (of the full Hamiltonian) with the correlated quantum numbers of $|n_1, n_2, n_3, n_5, n_6\rangle$. This labeling is just formal since due to resonances, they are no longer good quantum numbers or constants of motion for the levels of the full Hamiltonian. For convenience, we call $[n_1, n_2, n_3, n_5, n_6]$ the *formal quantum numbers*. (For the eigenstates of H_0, they are indeed the good quantum numbers.) Figure 5.2 shows part of the switching-off of the seven resonances. (Note that $2n_1 + n_2 + n_3 + 2n_5 + n_6 = 8$) Also shown are the labelings of the levels by $[n_1, n_2, n_3, n_5, n_6]$ and the diabatic correlation.

Now we can reconstruct the levels by the formal quantum numbers. Figure 5.1(a) is the reconstructed levels by $n_2 = 0, 1, \ldots, 8$. No apparent order is observed therein. The right most column in Fig. 5.1(b) shows the levels with $n_2 = 0$ and the rest are the reconstructions of these levels by $n_3 = 0, 1, \ldots, 8$. Similarly, the right most column in Fig. 5.1(c) shows the levels with $n_3 = 0$ $(n_2 = 0)$ and the rest are the reconstructions by $n_6 = 0, 2, 4, 6, 8$ in which each level is labeled by $[n_1 \, n_5]$, corresponding to the formal quantum numbers of the two C–H stretchings. Then, the regular pattern is obvious. For

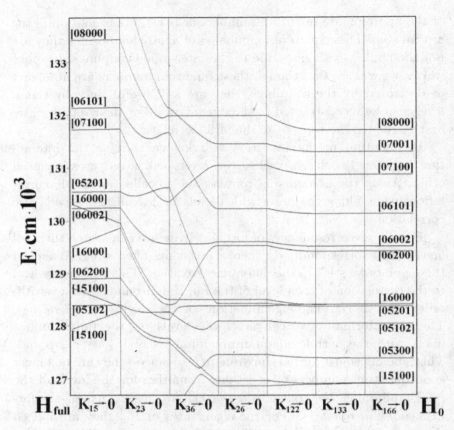

Fig. 5.2 The switching-off of the seven resonances of formaldehyde and the labelings of the levels by $[n_1, n_2, n_3, n_5, n_6]$ via the diabatic correlation. H_0 is the Hamiltonian without resonances. H_{full} is the full Hamiltonian.

each n_6 column, lower levels are almost degenerate, showing the local mode character. While for the higher levels, the nearest neighboring level spacings are almost the same, showing the character of the normal mode. These are the typical characters of a two-mode system (See Sections 2.3 and 2.4). In Fig. 5.1, only the level order by $n_2 = 0$, $n_3 = 0$ is shown. The others can also be reconstructed to show regular patterns in a similar way.

Above analysis shows that formal quantum numbers are indeed useful though they are not good quantum numbers or constants of motion in the strict sense due to inter-mode couplings. By the formal quantum numbers, levels which appear *irregularly* can be reconstructed to show regular patterns. This is very useful for the assignment and classification of highly excited vibrational levels.

5.3 Acetylene case

Acetylene possesses five in-plane motions. They are symmetric and antisymmetric stretchings of C–H (subscripts 1, 3), C–C stretching (subscript 2) and *trans*, *cis* bendings of C–H (subscripts 4,5). These modes have the following anharmonicities and resonances[3]:

$$\sum \omega_i n_i + \sum X_{ij} n_i n_j + 1/4 K_{4455}(a_4^+ a_4^+ a_5 a_5 + h.c.)$$

$$- 1/8 K_{3245}(a_3^+ a_2 a_4 a_5 + h.c.)$$

$$- 1/4 K_{1244}(a_1^+ a_2 a_4 a_4 + h.c.)$$

$$- 1/4 K_{1255}(a_1^+ a_2 a_5 a_5 + h.c.)$$

$$- 1/8 K_{1435}(a_1^+ a_4^+ a_3 a_5 + h.c.)$$

$$+ 1/4 K_{1133}(a_1^+ a_1^+ a_3 a_3 + h.c.).$$

The coefficients are tabulated in Table 5.2. For this Hamiltonian, the polyad number is $P = 5n_1 + 3n_2 + 5n_3 + n_4 + n_5$. This is easily checked by the orders of the creation and destruction operators in the coupling Hamiltonian. Figure 5.3 shows the switching-off of the resonances and the diabatic correlation from the full Hamiltonian to the zeroth-order Hamiltonian. By this procedure, we can label the levels of the full Hamiltonian with a set of formal quantum numbers. The right most column in Fig. 5.4 shows the 134 levels with $P = 15$. In the second right most column are the 51 levels with the formal quantum numbers $n_1 = n_3 = 0$. No apparent order is observed. The next step is the reconstruction of these 51 levels with $n_2 = 0, 1, \ldots, 5$. $n_2 = 0$ corresponds to pure C–H bending and levels are denoted by $[n_4, n_5]$. [15, 0] represents the *trans* bending which is lower than the *cis* bending [0,15]. Lower levels are almost degenerate while higher

Table 5.2   Coefficients (cm^{-1}) of the algebraic Hamiltonian for acetylene.

ω_1	3398.74
ω_2	1981.71
ω_3	3316.09
ω_4	609.02
ω_5	729.17
K_{4455}	−11.0
K_{3245}	−18.28
K_{1244}	6.38
K_{1255}	6.38
K_{1435}	29.04
K_{1133}	105.83
X_{11}	−26.57
X_{12}	−12.62
X_{13}	−105.09
X_{14}	−15.58
X_{15}	−10.85
X_{22}	−7.39
X_{23}	−6.10
X_{24}	−12.48
X_{25}	−1.57
X_{33}	−27.74
X_{34}	−6.96
X_{35}	−8.69
X_{44}	3.08
X_{45}	−2.41
X_{55}	−2.34

levels show almost equal level spacings. For nonzero n_2, the C–C stretching will affect the bending motion (like via the resonances K_{3245}, K_{1244}, K_{1255}). The level spacings are nearly the same.

The cases of formaldehyde and acetylene show that for the highly excited vibration, we can employ the quantum numbers of the zeroth-order Hamiltonian, called the formal quantum numbers which are retrieved through diabatic correlation, to reconstruct its levels in a very regular pattern. Although, due to resonances, these formal quantum numbers are no longer the constants of motion, they are still useful for the reconstruction of levels to show regularity. This means that formal quantum numbers are not completely meaningless quantities just because they are not exact.

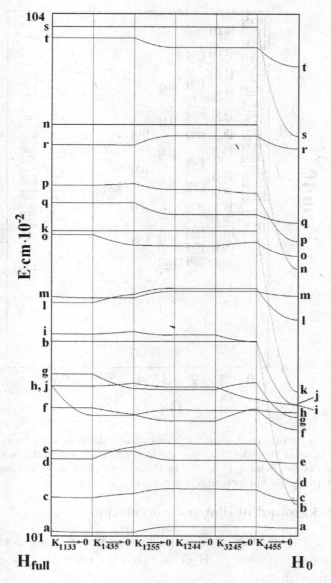

Fig. 5.3 The switching-off of the resonances of Hamiltonian H_{full} and the diabatic correlation to the zeroth-order Hamiltonian H_0 for acetylene.

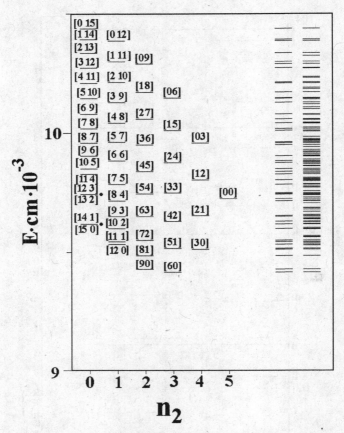

Fig. 5.4 The reconstruction of the levels of acetylene with $n_2 = 0, 1, .., 5$ by $[n_4, n_5]$. • denotes that the level spacing is very small. The right most column shows the 134 levels with $P = 15$. In the second right most column are the 51 levels with the formal quantum numbers $n_1 = n_3 = 0$.

5.4 Background of diabatic correlation

Hereby, we explain the physical background of diabatic correlation. The tri-atomic system of H_2O is taken as example. Its *algebraic* Hamiltonian is:

$$H = H_0 + H_{st} + H_F$$

$$H_0 = \omega_s(n_s + n_t + 1) + \omega_b \left(n_b + \frac{1}{2} \right)$$

$$+ X_{ss} \left[\left(n_s + \frac{1}{2} \right)^2 + \left(n_t + \frac{1}{2} \right)^2 \right]$$

$$+ X_{bb} \left(n_b + \frac{1}{2} \right)^2 + X_{st} \left(n_s + \frac{1}{2} \right) \left(n_t + \frac{1}{2} \right)$$

$$+ X_{sb}(n_s + n_t + 1) \left(n_b + \frac{1}{2} \right)$$

$$H_{st} = K_{st}(a_s^+ a_t + h.c.) + K_{DD}(a_s^+ a_s^+ a_t a_t + h.c.)$$
$$H_F = K_{sb}(a_s^+ a_b a_b + a_t^+ a_b a_b + h.c.)$$

where H_0 includes the nonlinear effects, H_{st} is the 1:1 and 2:2 resonances of s, t stretchings, H_F is the Fermi resonance among the bending (b) and stretchings (s, t). The coefficients are listed in Table 5.3[4] The polyad number is $P = n_s + n_t + n_b/2$.

(Note that this formulation is also adequate for D_2O and H_2S. Their coefficients are listed in Table 5.3. We will meet with the cases of D_2O and H_2S latter.)

Eigenstate ψ can be expressed as a linear combination of the eigenstates $|i_s j_t k_b\rangle$ of H_0 (That notations i_s, j_t, k_b are used instead of n_s, n_t, and n_b is just to stress that they are the quantum numbers for H_0.). Since $i_s + j_t + k_b/2$ is a constant, index k_b can be omitted. So we have

$$\psi = \sum C_{i_s j_t} |i_s j_t\rangle.$$

Table 5.3 The coefficients of the algebraic Hamiltonian. The unit is cm^{-1}.

	H_2O	D_2O	H_2S
ω_s	3890.64	2836.67	2735.15
ω_b	1645.25	1204.61	1229.83
X_{ss}	−82.06	−44.15	−48.69
X_{bb}	−16.18	−7.02	−9.57
X_{st}	−13.27	−10.57	−4.06
X_{sb}	−21.03	−13.14	−22.76
K_{st}	−42.80	−54.65	−8.11
K_{DD}	−0.19	−0.02	−0.58
K_F	−14.58	−5.83	−23.16

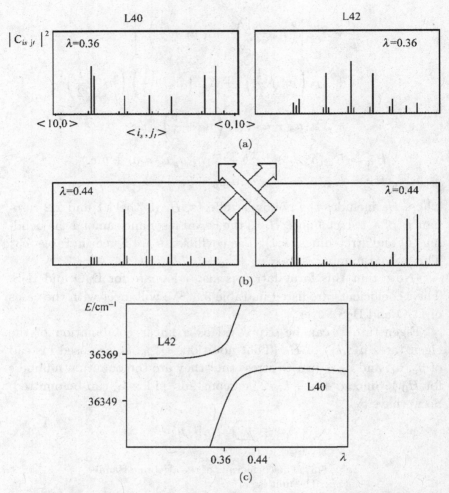

Fig. 5.5   (a) and (b) are the patterns of $|C_{i_s j_t}|^2$ of the 40th (L40) and 42nd (L42) levels of H_2O with $P = 10$, respectively. The Hamiltonian is $H_0 + H_{\text{st}} + \lambda H_{\text{F}}$ with $\lambda = 0.36, 0.44$. (c) shows the repulsion of these two levels as λ varies. However, if diabatic correlation is followed, the $|C_{i_s j_t}|^2$ patterns are preserved.

Figures 5.5(a) and (b) show $|C_{i_s j_t}|^2$ of the 40th (L40) and 42nd (L42) levels of H_2O with $P = 10$. The Hamiltonian is $H_0 + H_{\text{st}} + \lambda H_{\text{F}}$, with $\lambda = 0.36, 0.44$. The coordinate is expressed as $\langle i_s, j_t \rangle$. Due to interaction, these two levels repulse each other as shown in Fig. 5.5(c). From Figs. 5.5(a) and (b), it is seen that the distributions of $|C_{i_s j_t}|^2$ of these two levels change considerably as λ varies from 0.36 to 0.44. However, if the diabatic correlation is followed, the patterns of the distributions are preserved. This means that formal quantum numbers are related to the conservation phenomenon.

5.5 Approximately conserved quantum number

An alternative for us to classify vibrational manifolds is by the *approximate* quantum numbers. We can retrieve them by switching off H_{st} and H_{F} alternatively, in the following ways.

(1) Consider $H_0 + H_{\text{st}} + \lambda_2 H_{\text{F}}$, with λ_2 varying from 1 to 0. As $\lambda_2 = 0$, the eigenstates of $H_0 + H_{\text{st}}$ possess the exact quantum number n_{b}. As soon as the eigenstates of $H_0 + H_{\text{st}}$ are obtained, their n_{b}'s are known. By diabatic correlation, we can correlate the eigenstates of $H_0 + H_{\text{st}} + H_{\text{F}}$ with those of $H_0 + H_{\text{st}}$. Since the effect of H_{F} is small, n_{b} is the approximate quantum number of the full Hamiltonian ($H_0 + H_{\text{st}} + H_{\text{F}}$).

(2) Consider $H_0 + \lambda_1 H_{\text{st}} + H_{\text{F}}$, with λ_1 varying from 1 to 0. As $\lambda_1 = 0$, $H_0 + H_{\text{F}}$ possesses the approximate quantum number:

$$\sum |C_{i_s j_t}|^2 |i_s - j_t|.$$

For convenience, we denote it as $|n_s - n_t|$. We have this approximate constant of motion because $(a_s^+ a_b a_b + a_t^+ a_b a_b + h.c.)$ increases or decreases the actions of s and t simultaneously in a form of linear combination. By the diabatic correlation, we can label $|n_s - n_t|$ on the eigenstates of the full Hamiltonian. They are the approximate constants. (The effect of H_{st} is smaller in $H_0 + H_{\text{st}} + H_{\text{F}}$).

From $|n_s - n_t|$ and n_{b} ($=2(P - (n_s + n_t))$), we can calculate (n_s, n_t). They are the approximately conserved quantities of the eigenstates of $H_0 + H_{\text{st}} + H_{\text{F}}$. Table 5.4 shows $|n_s - n_t|$, $n_s + n_t$ and (n_s, n_t) of the 36 levels of D_2O with $P = 7$. Listed in the Table are also their formal quantum numbers $[n_s, n_t]$. Obviously, the truncated (n_s, n_t) are identical to $[n_s, n_t]$.

Table 5.4   The approximately conserved $|n_s - n_t|$, $n_s + n_t$ and (n_s, n_t) of the 36 levels of D_2O with $P = 7$ and their formal quantum numbers $[n_s, n_t]$.

Level	$|n_s - n_t|$	$n_s + n_t$	(n_s, n_t)	$[n_s, n_t]$
1	0	0	(0, 0)	[0 0]
2	0.98	1	(0.99, 0.04)	[1 0]
3	1.04	1	(1.02, −0.02)	[1 0]
4	2.00	2	(2.00, 0.00)	[2 0]
5	2.01	2	(2.00, −0.01)	[2 0]
6	0	2	(1.00, 1.00)	[1 1]
7–8	3.02	3	(3.01, −0.01)	[3 0]
9	0.96	3	(1.98, 1.02)	[2 1]
10–11	4.03	4	(4.02, −0.02)	[4 0]
12	1.05	3	(2.03, 0.98)	[2 1]
13	2.01	4	(3.01, 1.00)	[3 1]
14–15	5.05	5	(5.03, −0.03)	[5 0]
16	1.99	4	(3.00,1.01)	[3 1]
17–18	6.33	6	(6.17, −0.17)	[6 0]
19–20	6.5	7	(6.75,0.25)	[7 0]
21	3.00	5	(4.00,1.00)	[4 1]
22	0	4	(2.00,2.00)	[2 2]
23	3.00	5	(4.00,1.00)	[4 1]
24–25	3.99	6	(5.00,1.01)	[5 1]
26	0.97	5	(2.99,2.02)	[3 2]
27–28	4.96	7	(5.98,1.02)	[6 1]
29	1.99	6	(4.00,2.00)	[4 2]
30	1.03	5	(3.02,1.99)	[3 2]
31	2.99	7	(5.00,2.01)	[5 2]
32	1.98	6	(3.99,2.01)	[4 2]
33	2.99	7	(5.00,2.01)	[5 2]
34	0	6	(3.00,3.00)	[3 3]
35	0.99	7	(4.00,3.01)	[4 3]
36	1	7	(4.00,3.00)	[4 3]

Figure 5.6(a) shows the reconstruction of the 36 levels of D_2O with $P = 7$ by $[n_s, n_t]$. Figures 5.6(b) and (c) show the reconstructions of H_2O and H_2S cases. From these, we conclude:

(1) The energy levels of H_2O, D_2O, H_2S are different. However, from the reconstructed level diagrams, they share the same *pattern*.

(2) For each n_b, there is a series of states labeled by $[n_s, n_t]$. Lower levels have different n_s and n_t. Higher levels have closer n_s and

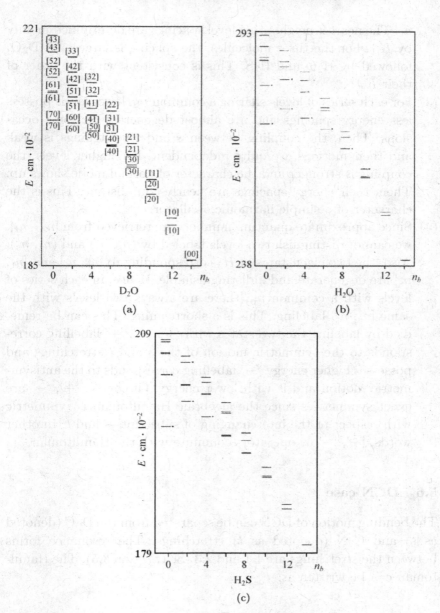

Fig. 5.6   (a) The reconstruction of the 36 levels of D_2O with $P = 7$ by $[n_s, n_t]$.
(b),(c) are those of H_2O and H_2S. • shows that energy spacing is small.

n_t. The nearest neighboring level spacings are determined mainly by K_{st}. For the three molecules, the spacing is largest for D_2O, followed by H_2O and H_2S. This is consistent with the order of their K_{st}.

(3) For each series of levels sharing a common n_b, lower levels possess less energy spacings and are almost degenerate on some occasions. Then, the coupling between s and t stretchings is weak and their motions are quite independent. For higher levels, the coupling is stronger and the character of normal mode shows up. Then, their energy spacings are nearly equi-distant. This is the character of a simple harmonic oscillator.

(4) Since approximate quantum numbers are retrieved from $|n_s - n_t|$, we cannot distinguish two levels labeled by (n_s, n_t) and (n_t, n_s). Also, the two eigenstates of H_0 corresponding to $[n_s, n_t]$ and $[n_t, n_s]$ are degenerate and indistinguishable. Hence, in each series of levels with a common n_b, there are always two levels with the same $[n_s, n_t]$ labeling. This is a shortcoming. This can be remedied by labeling the two states with '+', '−'. '+' labelling corresponds to the symmetric motion of the s and t stretchings and possesses higher energy. '−' labelling corresponds to the antisymmetric motion and is with lower energy. Of course, '+', '−' are exact symmetries since the algebraic Hamiltonian is symmetric with respect to the interchanging of subscripts s and t. In other words, '+', '−' as operators commute with the Hamiltonian.

5.6   DCN case

The bending motion of DCN can be separated from the D–C (denoted as s) and C–N (denoted as t) stretchings. The resonance forms between the stretchings are 1:1 and 2:3 (See Section 3.5). The Hamiltonian can be written as:

$$H = H_0 + H_{st} + H_K$$
$$H_0 = \omega_s \left(n_s + \frac{1}{2} \right) + \omega_t \left(n_t + \frac{1}{2} \right)$$

$$+ X_{ss} \left(n_s + \frac{1}{2} \right)^2 + X_{tt} \left(n_t + \frac{1}{2} \right)^2$$

$$+ X_{st} \left(n_s + \frac{1}{2} \right) \left(n_t + \frac{1}{2} \right)$$

$$H_{st} = K_{st}(a_s^+ a_t + h.c.)$$

$$H_K = K(a_s^+ a_s^+ a_t a_t a_t + h.c.).$$

The coefficients are listed in Table 3.1.

H_0 will not destroy n_s, n_t. For H_{st}, $P_1 = n_s + n_t$ is conserved. For H_K, $P_2 = n_s/2 + n_t/3$ is conserved. For H, except energy, there are no constants of motion. This is a *simple* case but it possesses very *complicated* dynamics. For its dynamical details, see Chapter 3.

By switching off H_K and the diabatic correlation, we have the approximate constant of motion P_1:

$$P_1 = \sum (i_s + j_t)|C_{i_s j_t}|^2.$$

Similarly, if H_{st} is switched off, we have another approximate constant P_2:

$$P_2 = \sum (i_s/2 + j_t/3)|C_{i_s j_t}|^2.$$

From P_1, P_2, the approximate quantum numbers, (n_s, n_t), can be obtained. For the 66 levels of DCN, they are tabulated in Table 5.5. Quite surprisingly, all (n_s, n_t) are integers. This shows that they are not seriously destroyed. (n_s, n_t) are exactly the formal quantum numbers in this case. We can reconstruct these 66 levels by P_1 as shown in Fig. 5.7. The right most column is the 66 levels. No regularity is obvious. While for each P_1, the levels are labeled by $(n_s, n_t) = (0, P_1), (1, P_1 - 1), \ldots, (P_1, 0)$, consecutively. They show almost equal energy spacings. (With careful scrutiny, the spacings for the levels in the middle of a column are smaller. This corresponds to the separatrix. For details, see Section 2.3.) Therefore, by approximate quantum numbers, the levels seems in disorder originally, can be reconstructed into a very regular pattern.

In conclusion, from the cases of formaldehyde, acetylene, H_2O, D_2O, H_2S and DCN, it is known that due to resonances, most

Table 5.5   The approximate $P_1 = n_s + n_t$, $P_2 = n_s/2 + n_t/3$
and (n_s, n_t) for the 66 levels of DCN.

Level	(P_1, P_2)	(n_s, n_t)
1	(0, 0)	(0, 0)
2	(1, 0.333)	(0,1)
3	(1, 0.5)	(1, 0)
4	(2, 0.666)	(0, 2)
5	(2, 0.833)	(1, 1)
6	(2, 0.999)	(2, 0)
7	(3, 0.999)	(0, 3)
8	(3, 1.166)	(1, 2)
9	(3, 1.333)	(2, 1)
10	(4, 1.333)	(0, 4)
11	(3, 1.5)	(3, 0)
12	(4, 1.5)	(1, 3)
13	(4, 1.666)	(2, 2)
14	(5, 1.666)	(0, 5)
15	(4, 1.833)	(3, 1)
16	(5, 1.833)	(1, 4)
17	(4, 2)	(4, 0)
18	(5, 2)	(2, 3)
19	(5, 2.166)	(3, 2)
20	(6, 2)	(0, 6)
21	(5, 2.333)	(4, 1)
22	(6, 2.166)	(1, 5)
23	(6, 2.333)	(2, 4)
24	(5, 2.5)	(5, 0)
25	(6, 2.5)	(3, 3)
26	(7, 2.333)	(0, 7)
27	(6, 2.666)	(4, 2)
28	(7, 2.499)	(1, 6)
29	(6, 2.833)	(5, 1)
30	(7, 2.666)	(2, 5)
31	(7, 2.833)	(3, 4)
32	(6, 3)	(6,0)
33	(7, 3)	(4, 3)
34	(8, 2.666)	(0, 8)
35	(7, 3.166)	(5, 2)
36	(8, 2.833)	(1, 7)
37	(8, 2.999)	(2, 6)

(*Continued*)

Table 5.5 (*Continued*)

Level	(P_1, P_2)	(n_s, n_t)
38	(7, 3.333)	(6, 1)
39	(8, 3.166)	(3, 5)
40	(8, 3.333)	(4, 4)
41	(7, 3.5)	(7, 0)
42	(9, 2.999)	(0, 9)
43	(8, 3.499)	(5, 3)
44	(9, 3.166)	(1, 8)
45	(8, 3.666)	(6, 2)
46	(9, 3.333)	(2, 7)
47	(9, 3.5)	(3, 6)
48	(8, 3.833)	(7, 1)
49	(9, 3.666)	(4, 5)
50	(10, 3.333)	(0, 10)
51	(9, 3.833)	(5, 4)
52	(8, 4)	(8, 0)
53	(10, 3.5)	(1, 9)
54	(9, 4)	(6, 3)
55	(9, 4.166)	(7, 2)
56	(10, 3.666)	(2, 8)
57	(10, 3.833)	(3, 7)
58	(9, 4.333)	(8, 1)
59	(10, 4)	(4, 6)
60	(10, 4.166)	(5, 5)
61	(9, 4.5)	(9, 0)
62	(10, 4.333)	(6, 4)
63	(10, 4.5)	(7, 3)
64	(10, 4.666)	(8, 2)
65	(10, 4.833)	(9, 1)
66	(10, 5)	(10, 0)

conserved quantities are destroyed in the highly excited vibration. However, there are approximate constants of motion by which levels can be reconstructed into a very regular pattern. From the reconstructed diagram, dynamics concerning resonances and the origin of approximate quantum numbers become transparent. For molecular highly excited vibrational levels, though the number is immense and complicated, they are assignable and can be classified.

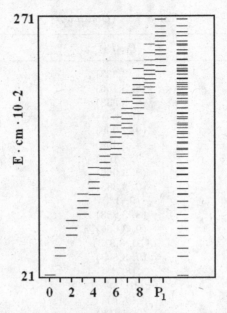

Fig. 5.7   The reconstruction of the levels of DCN by P_1. For each P_1, the levels are labeled by $(n_s, n_t) = (0, P_1), (1, P_1 - 1), \ldots, (P_1, 0)$. The right most column has the 66 levels which possess no obvious regularity.

5.7   Density ρ in the coset space

The Hamiltonians in the above cases can be cast to the expressions in the coset space. Then, we have H expressed as $H\,(q_s, p_s, q_t, p_t)$. (q_s, p_s, q_t, p_t) are the coordinates of the coset spaces. (For details, see Chapter 2.)

Given an eigenenergy E, from $E = H(q_s, p_s, q_t, p_t)$, we can obtain the (q_s, p_s, q_t, p_t) solution space. As long as the number of solutions is large, we can construct the density ρ in the coset space (or phase space).

An eigenstate is a linear combination of $\{|i_s, j_t\rangle\}$. This is the concept of quantum mechanics. In the phase space, the dynamical property is realized by ρ. For each $\langle i_s, j_t \rangle$ with integral i_s and j_t, there are the corresponding (q_s, p_s, q_t, p_t) points. The correlation of multiple (q_s, p_s, q_t, p_t) points to a given $\langle i_s, j_t \rangle$ is anticipated. Of course, there are some $\langle i_s, j_t \rangle$ with which no (q_s, p_s, q_t, p_t) points can be associated. For brevity, the density $\rho(\langle i_s, j_t \rangle)$ corresponding to $\langle i_s, j_t \rangle$ is shortened to $\rho(i_s, j_t)$.

The algorithm for calculating $\rho(i_s, j_t)$ is to mesh the whole phase space into 1.6×10^9 grids (q_α, p_α coordinates are partitioned into 200 grids, respectively.). For the given i_s, j_t, we calculate the solutions satisfying the following conditions (with H_2O as an example):

$$|H(q_\alpha, p_\alpha) - E| < 0.1 (\text{cm}^{-1})$$
$$|(q_s^2 + p_s^2)/2 - i_s| < 0.1$$
$$|(q_t^2 + p_t^2)/2 - j_t| < 0.1.$$

The number of the solutions can then be assigned as $\rho(i_s, j_t)$. For the formal quantum numbers $[n_s, n_t]$, the corresponding $\rho([n_s, n_t])$ can also be calculated.

The calculation shows that as P is small, the correlation between $\rho(i_s, j_t)$ and $|C_{i_s j_t}|^2$ is not good. As P is large, the correlation becomes good. Shown in Fig. 5.8 is the case of the 110th level (denoted as L110) of H_2O, $P = 15$. The correlation is apparent. Figure 5.9 shows the case of L40 of DCN. The correlation between $\rho(i_s, j_t)$ and $|C_{i_s j_t}|^2$ is also apparent. We note that ρ is a classical quantity, $|C_{i_s j_t}|^2$ is the concept of quantum mechanics. They are calculated independently. Their consistency is just the Correspondence Principle! In

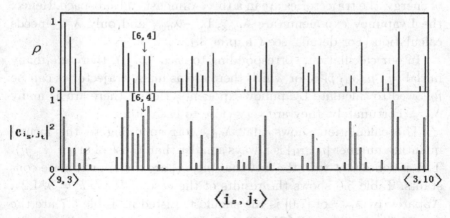

Fig. 5.8   The correlation between $\rho(i_s, j_t)$ and $|C_{i_s j_t}|^2$ for the 110th level of H_2O, $P = 15$.

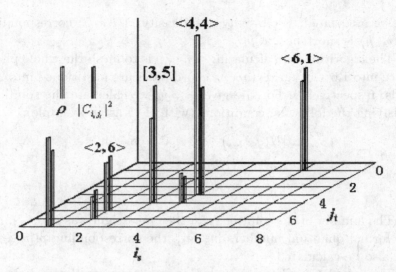

Fig. 5.9   The correlation between $\rho(i_s, j_t)$ and $|C_{i_s j_t}|^2$ for the 40th level of DCN, $P = 10$.

these two figures, $\rho([n_s, n_t])$ are also shown. It seems that it is a rule that $\rho([n_s, n_t])$ is always distinct.

5.8   Lyapunov exponent analysis

For our four-dimensional (q_s, p_s, q_t, p_t) space, due to the conservation of energy, the trajectories are in a three-dimensional subspace. Hence, the Lyapunov exponents are $\lambda_{\max}$, 0, $-\lambda_{\max}$ and only $\lambda_{\max}$ needs calculation (For details, see Chapter 3.).

In our calculation, corresponding to each $\langle i_s, j_t \rangle$, there are many initial $(q_s, p_s, q_t, p_t)_0$ for which there are as many trajectories can be followed to calculate Lyapunov exponents. Hence, there are as many $\lambda_{\max}$. Fortunately, they are very close to each other.

The calculation shows that $\lambda_{\max}$ corresponding to the formal quantum number $[n_s, n_t]$ is always smaller than those of other $\langle i_s, j_t \rangle$. This is consistent with the fact that $[n_s, n_t]$ are the approximate constants. Table 5.6 shows the results of the case of H_2O, $P = 9$, L24. Apparently, $\lambda_{\max}$ of $[7,0]$ is the smallest. Listed in Table 5.7 are the results for the case of DCN L20, L25, L30. $\lambda_{\max}$ corresponding to the formal quantum numbers are always the smallest.

Table 5.6   The Lyapunov exponents for H_2O, $P = 9$, L24.

$\langle i_s, j_t \rangle$	$\lambda_{\max}$ (ps^{-1})
[7,0]	0.10
$\langle 5, 0 \rangle$	0.11
$\langle 6, 0 \rangle$	0.13
$\langle 3, 2 \rangle$	0.15
$\langle 2, 2 \rangle$	0.17
$\langle 8, 1 \rangle$	0.19
$\langle 6, 1 \rangle$	10.05
$\langle 4, 1 \rangle$	10.21
$\langle 5, 1 \rangle$	10.28
$\langle 3, 1 \rangle$	10.86

Table 5.7   The Lyapunov exponents for DCN, L20, L25, L30.

Level	$\langle i_s, j_t \rangle$	$\lambda_{\max}$ (ps^{-1})
	[0,6]	0.04
20	$\langle 2, 3 \rangle$	3.01
	$\langle 3, 2 \rangle$	4.91
	$\langle 4, 1 \rangle$	3.57
	[3,3]	0.06
25	$\langle 4, 2 \rangle$	4.38
	$\langle 2, 4 \rangle$	7.92
	[2,5]	0.59
	$\langle 3, 4 \rangle$	11.40
30	$\langle 4, 2 \rangle$	13.21
	$\langle 5, 1 \rangle$	8.99
	$\langle 1, 6 \rangle$	9.83
	$\langle 4, 3 \rangle$	9.60

To demonstrate that the trajectories whose initial points in the phase space corresponding to the formal quantum number are more regular and possess the least Lyapunov exponent, the Poincare surface (q_s, p_s) $(p_t < 0$, along $q_t = 0)$ structures of the trajectories originating from [3, 3] and $\langle 4, 2 \rangle$ for the case of DCN, L25 are shown in Figs. 5.10(a), (b). Indeed, the trajectories originating from [3, 3] are quite regular while those from $\langle 4, 2 \rangle$ are chaotic.

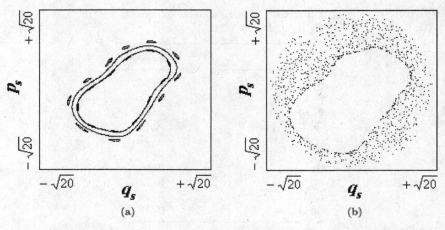

Fig. 5.10   The Poincare surface (q_s, p_s) $(p_t < 0$, along $q_t = 0)$ structures of the trajectories originating from $[3, 3]$ (a) and $\langle 4, 2 \rangle$ (b) for the case of DCN, L25.

References

1. Gray S K, Davis M J. *J. Chem. Phys.*, 1989, 90: 5420.
2. Reisner D E, Field R W, Kinsey J L, Dai H L. *J. Chem. Phys.*, 1984, 80: 5968.
3. Solina S A B, O'Brien J P, Field R W, Polik W F. *J. Phys. Chem.*, 1996, 100: 7797.
4. Iachello F, Oss S. *J. Mol. Spectrosc.*, 1990, 142: 85.

Chapter 6

Dixon Dip

6.1 Significance of level spacings

Vibrational system energy is quantized. From the spectroscopic viewpoint, it is the level spacings that we can observe. For a simple harmonic oscillator, the level spacing is equi-distant and equal to its *classical* frequency (or angular velocity), i.e., $\Delta E/(h/2\pi) = \omega$. Hence, the significance of the level spacings is that they offer the *classical* frequencies/angular velocities of an oscillator (For Morse oscillator and pendulum, this quantity is energy or level dependent).

By this interpretation, we recognize that Heisenberg's energy matrix, in which the off-diagonal terms are the level spacings, offers the *full* information of a quantized system. In Heisenberg's view, a quantized system is a collection of many oscillators whose *classical* frequencies/angular velocities are associated with the (quantized) level spacings. In contrast, Schroedinger's approach to a quantum system is not so informative.

6.2 Dixon dip

As known, a resonance is a mimic of the pendulum motion for which there are two dynamical realms separated by the *separatrix* and two associated stable and unstable fixed points (See Sections 1.4 and 2.3). This is a classical viewpoint. Classically, a quantum level (state) can be viewed as a subspace in the dynamical phase space. Therefore, levels can be classified as those lying below and above the separatrix in the energy scale due to resonance. For a pendulum, its oscillating frequency is expected to become smaller as the motion is close to the separatrix due to the nonlinear effect. Quantum mechanically, this will be reflected in that the nearest neighboring level spacing for those levels around the separatrix will reach a minimum. This is the so-called Dixon dip.[1-4] This concept is integrated by the quantal and classical analogues. We note that Dixon dip appears not only in the levels sharing a well-defined polyad number, but also when the resonance is very seriously perturbed by other complicated interactions, like in the systems of Henon–Heiles and quartic potentials of which resonances are hardly well defined. For the molecular systems like H_2O and DCN possessing multiple resonances, Dixon dip will be destroyed by the overlapping of resonances which, as conjectured by Chirikov, will lead to chaos. This will be further analyzed by an independent Lyapunov exponent analysis as shown below.

6.3   Dixon dips in the systems of Henon–Heiles and quartic potentials

The Henon–Heiles system is a two-dimensional harmonic oscillator with perturbing potential of the form

$$(p_x^2 + p_y^2 + x^2 + y^2)/2 + \lambda x(y^2 - x^2/3)$$

(λ is chosen as 0.05 in our calculation.) For this system, it is straightforward to cast the Hamiltonian in the second quantized language via

$$p_\alpha = i(a_\alpha^+ - a_\alpha)/\sqrt{2}$$
$$\alpha = (a_\alpha^+ + a_\alpha)/\sqrt{2} \qquad (\alpha = x, y)$$

for which a_α^+ and a_α are the creation and destruction operators, respectively. The perturbing potential is very complicated in terms of the second quantized operators. However, embedded therein, there is one term that is of resonance type, i.e., $a_x^+ a_y a_y + h.c.$ for which

$$P = n_x + n_y/2$$

is conserved. Here, n_x and n_y are the actions on the x and y coordinates. Since there are other perturbations, P is not well defined. For this system, it is simple to construct the matrix Hamiltonian by employing the bases $|n_x, n_y\rangle$ with n_x, n_y ranging from 0 to n which is chosen by arbitration. The eigenenergies are then obtained as the matrix Hamiltonian is diagonalized. By reducing λ to zero, we can diabatically assign the quantum numbers of the zeroth-order states to those under the full perturbing potential. Hence, each level can be assigned by an approximate constant of motion P. Figure 6.1(a) shows the nearest neighboring level spacings for the levels possessing a common P up to 7. The dips are obvious. (For clarity, only some of the cases are shown. The numbers on the coordinate are the numbering of the nearest level spacings among the levels. This is the same for other figures.)

The quartic system has potential $\alpha x^2 y^2$ ($\alpha = 0.25$ in our calculation.) The Hamiltonian is

$$(p_x^2 + p_y^2)/2 + \alpha x^2 y^2.$$

By second quantization, it can be seen that embedded in the quartic potential there is the Darling–Dennison (2:2) resonance:

$$a_x^+ a_x^+ a_y a_y + h.c.$$

Analogously, each eigenstate can be assigned by an approximate constant of motion

$$P = n_x + n_y.$$

Shown in Fig. 6.1(b) are the nearest neighboring level spacings among the levels belonging to $P = 7$ to 11. The dips are distinct.

In summary, it should be stressed that these results are not trivial if one notes that P is only approximate under the other perturbing potentials.

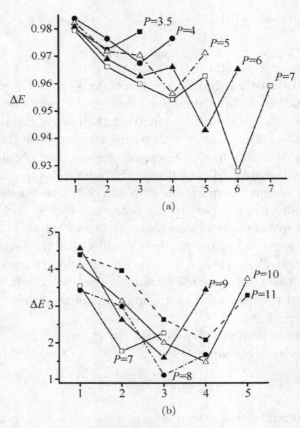

Fig. 6.1   (a) The nearest neighboring level spacings for the levels possessing a common P up to 7 for the system of Henon–Heiles potential. (For clarity, only some of the cases are shown.) (b) The nearest neighboring level spacings among the levels belonging to $P = 7$ to 11 for the system of quartic potential. The numbers on the coordinate are the numbering of the nearest neighboring level spacings among the levels.

6.4   Destruction of Dixon dip under multiple resonances

A. H_2O system

For the system of H_2O, we consider two resonances. One is between the two equivalent O–H stretchings (with subscripts s and t) which is 1:1 resonance. The other is the Fermi resonance between the bending (with subscript b) and the two stretchings. In second quantization, they are:

$$a_s^+ a_t + h.c.$$

and

$$a_s^+ a_b a_b + a_t^+ a_b a_b + h.c.$$

Four approximate polyad numbers can be defined for this system. They are $P_1 = n_s + n_t$, $P_2 = |n_s - n_t|$, $P_3 = n_s + n_b/2$ (this is the same as $n_t + n_b/2$) and $P_4 = n_s + n_t + n_b/2$ with n_s, n_t and n_b the actions on the stretchings and bending. P_1 is due to the 1:1 resonance. P_2 is due to the linear combination of $a_s^+ a_b a_b + a_t^+ a_b a_b + h.c.$. P_3 is due to the Fermi resonance. P_4 is due to both the 1:1 and Fermi resonances. P_4 is strict under both resonances.

For the case of P_1, the dips are not sharp. This can be due to that the 1:1 resonance is not strong. Shown in Fig. 6.2(a) is the case of $P_1 = 11$. Note that due to the equivalence of the two O–H stretchings, there are almost degenerate states. They are the local modes in which the coupling between the two O–H stretchings is small so that they behave more or less independently. This results in zero level spacing in Fig. 6.2(a). For the case of P_2, multiple dips are apparent. This is shown in Fig. 6.2(b) for $P_2 = 2$. This can be attributed to the more complicated Fermi resonance. (It contains two terms and their Hermitian conjugates: $a_s^+ a_b a_b + a_t^+ a_b a_b + h.c.$) For the case of P_3, dips appear in certain situations with smaller P_3. This is shown in Fig. 6.2(c) for $P_3 = 3$. We note that due to multiple resonances, P_1, P_2 and P_3 are not well defined. However, the existence of the Dixon dips is still obvious in some cases. However, the destruction of Dixon dip is also obvious due to multiple resonances. This will be further explored in the DCN system in the next section. P_4 is strictly conserved under both 1:1 and Fermi resonances. However, the coexistence of these two resonances can cause the dip appearance in a complicated manner. The case of $P_4 = 6$ is shown in Fig. 6.2(d). Indeed, there is an obvious concave structure in the level spacings (It may be also viewed as two dips). For the cases with larger P_4, the nearest neighboring level spacings show two distinct dips with zigzag

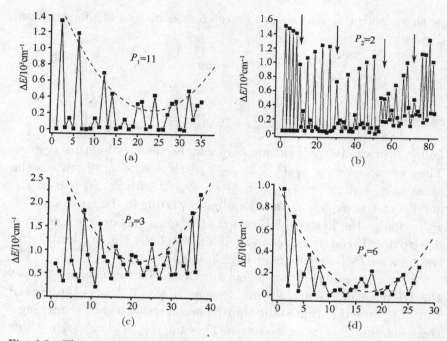

Fig. 6.2   The nearest neighboring level spacings for the levels of H_2O. Dashed lines and arrows are drawn for viewing the dips.

structure (not shown). Apparently, this is due to the coexistence of 1:1 and Fermi resonances.

B. DCN system: Overlapping of resonances and chaos

We consider the DCN system as composed of two Morse stretchings with 1:1 and 2:3 resonances. The consideration of 2:3 resonance is based on that the stretching frequency ratio of D–C to C–N (their frequencies are around 2681 cm^{-1} and 1949 cm^{-1}, respectively) is around 1.38, which is between 1 and 1.5. For this system, the approximate polyad numbers are:

$$P_1 = n_s + n_t$$

and

$$P_2 = n_s/2 + n_t/3$$

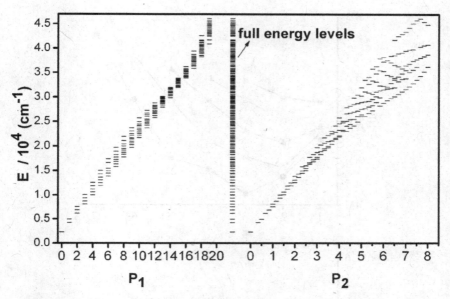

Fig. 6.3 The levels of the two stretchings of DCN up to 45,000 cm^{-1} and their unzipping by $P_1 = n_s + n_t$ and $P_2 = n_s/2 + n_t/3$.

n_s and n_t are the actions on the D–C and C–N stretchings, respectively. Shown in Fig. 6.3 are the levels up to 45,000 cm^{-1}, which is close to dissociation and their unzipping by P_1 and P_2. (The calculation is based on the Morse model for the two stretchings with 1:1 and 2:3 resonances. See Section 3.5).

For $P_1 < 7$ (i.e., for those levels <20,000 cm^{-1}), the nearest neighboring level spacings show simple dips. (See Fig. 6.4(a)). For higher levels (associated with $P_1 > 7$), the dip phenomenon is no longer evident and the nearest neighboring level spacings show a zigzag structure. This means then that the 1:1 resonance is seriously perturbed by the 2:3 resonance. For $P_2 < 5$ (levels <30,000 cm^{-1}) the dip phenomenon is not evident, while for larger P_2 (with level energy up to 45,000 cm^{-1}), the dips are distinct again. This is shown in Fig. 6.4(b). By this comparison, we may conclude that for those levels in between 20,000 cm^{-1} and 30,000 cm^{-1}, the two resonances are seriously mutually perturbed and chaos can predominate. For the

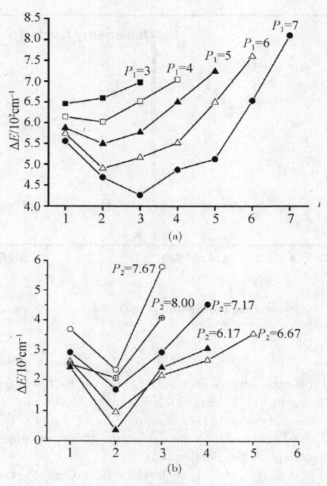

Fig. 6.4 The nearest neighboring level spacings of DCN for the levels with (a) $P_1 < 7$; (b) $P_2 > 5$.

lower and higher levels, either 1:1 or 2:3 resonance is operative and levels are relatively less chaotic, dynamically.

For showing the degree of chaos, the averaged largest Lyapunov exponent, $\langle\lambda\rangle$ is calculated by randomly choosing 200 initial points in the dynamical space corresponding to a level. The result is shown in Fig. 6.5. Clearly, as the level is above 10,000 cm^{-1}, its dynamics is becoming chaotic. In this energy region, the dynamics is mainly

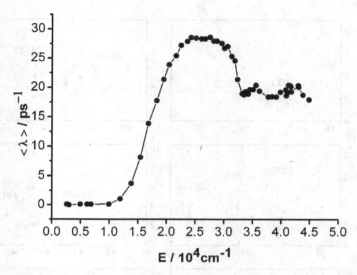

Fig. 6.5 The averaged largest Lyapunov exponent $\langle\lambda\rangle$ as a function of the level energy.

by the 1:1 resonance, perturbed slightly by the 2:3 resonance. The dynamics reaches the highest degree of chaos as the level energy is 25,000 cm^{-1}. In between 20,000 cm^{-1} and 30,000 cm^{-1}, both 1:1 and 2:3 resonances are operative. Their overlapping, as conjectured by Chirikov, leads to chaos. For the levels above 30,000 cm^{-1}, the diminishing of the 1:1 resonance leaves only the 2:3 resonance prominent. Meanwhile, the overlapping of the two resonances and the Lyapunov exponent are also decreasing. Hence, the degree of chaos drops slightly even as the level energy increases more!

Shown in Fig. 6.6 are the Dixon dips as a function of level energy. Dixon dips show the behavior of the levels as they cross over the separatrix.

In summary, our work shows that this dip phenomenon is so evident even for a resonance that is seriously perturbed by other interactions like in the systems of Henon–Heiles and quartic potentials for which the polyad numbers are not well-defined and only approximate. This demonstrates a *global* viewpoint for unfolding the dynamics which does not rely on individual levels. In general, the dip

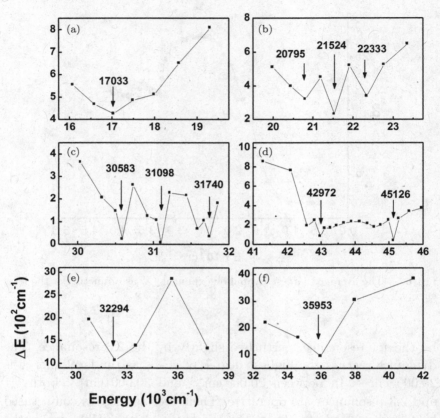

Fig. 6.6   The Dixon dip as a function of level energy. (a) $P_1 = 7$; (b) $P_1 = 9$; (c) $P_1 = 14$; (d) $P_1 = 19$; (e) $P_2 = 6.333$; (f) $P_2 = 7$.

phenomenon is distinct for a simple resonance. The emerging of multiple resonances definitely will complicate the level structure and may make the dip phenomenon less obvious. We demonstrated this for the system of H_2O. For the system of DCN, the dips are only apparent for the low and high levels while not for those levels in between. By Chirikov's conjecture, this is due to the overlapping of resonances which will lead to chaos. Independently, this phenomenon is analyzed in terms of the degree of chaos by the averaged Lyapunov exponent. These two results show consistent interpretations. This algorithm is of potential value in studying the dynamical properties of the highly excited vibrational system, including the *transition state*, of which multiple resonances are co-existent, and therefore, chaos is prevailing.

6.5 Dixon dip and chaos

Shown in Fig. 6.7 are the Dixon dips with their averaged largest Lyapunov exponents for DCN. Indeed, these Lyapunov exponents

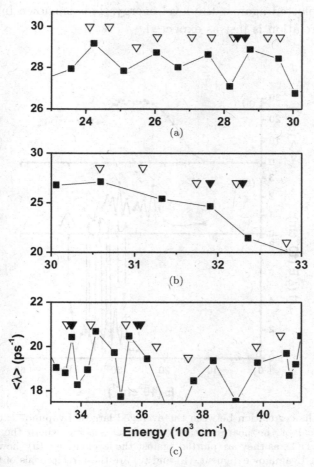

Fig. 6.7 The Dixon dips with their averaged largest Lyapunov exponents for DCN. ∇ for those by P_1, $\blacktriangledown$ for those by P_2. Black squares are for those of general level energies.

are larger than their nearby ones. This confirms the close correlation between the dip and chaos. We note that this consistency comes from two independent viewpoints, one is from the behavior of the level spacing as it crosses over the separatrix and the other is based on the classical dynamical interpretation of the energy level.

Based on the above assertion, we may expect correlation between the averaged largest Lyapunov exponent and the reciprocal of the nearest neighboring level spacing where the Dixon dip appears as they are plotted against the level energy. This is shown in Fig. 6.8. Their correlation is just as expected.

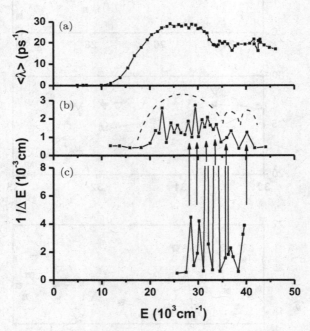

Fig. 6.8   The correlation between the averaged largest Lyapunov exponent and the reciprocal of the nearest neighboring level spacing where the Dixon dip appears for DCN as they are plotted against the level energy. (a) shows the averaged largest Lyapunov exponent, (b) and (c) are those reciprocals of the nearest neighboring level spacings classified by P_1 and P_2, respectively. The dashed curve in (b) is to show the profile. The arrows between (b) and (c) show the level energy correspondence.

In conclusion, we note that Dixon dip, which is based on the quantized level spacing, and the Lyapunov exponent, which is based on the classical dynamics, do show complementary and consistent information. The former idea is *global*, while the latter is *local*. However, both serve our understanding of the vibrational dynamics.

References

1. Dixon R N. *Trans. Farad. Soc.*, 1964, 60: 1363.
2. Child M S. *J. Mol. Spectrosc.*, 2001, 210: 157.
3. Yang S, Tyng V, Kellman M E. *J. Phys. Chem.*, 2003, A 107: 8345.
4. Svitak J, Li Z, Rose J, Kellman M E. *J. Chem. Phys.*, 1995, 102: 4340.

Chapter 7

Quantization by Lyapunov Exponent and Periodic Trajectories

7.1 Introduction

Bohr first pointed out the idea of quantization in a Coulomb field. Sommerfeld then extended the idea to a system: if the space coordinate q is a cyclic variable then the integration of its momentum p by space coordinate will be a multiple of Planck's constant h. Later, Einstein, Brillouin and Keller (EBK) gave the quantization condition of an integrable system as:

$$\oint pdq = h(n + \delta).$$

The integral is called the action. δ is the Maslov index. n is an integer. For vibration and electronic motion, $\delta = 1/2$. For in-plane rotation, $\delta = 0$.

Gutzwiller has pointed out that under semi-classical approximation, the quantal density of states can be related to the periodic trajectories. Gutzwiller's formula involves the convergence of an infinite series. The complete exploitation of periodic trajectories is not an easy task. Indeed, there is difficulty in the application of Gutzwiller's trace formula.[1] The essence of the issue is the long-time unsolved problem: the semi-classical quantization of nonintegrable or chaotic systems.

In this chapter, we will address how, via the analysis of the Lyapunov exponent, this problem can be partially solved or that light can be shed on the issue in certain cases. It should be emphasized that our choice of trajectories is random and our quantization is by the least averaged Lyapunov exponent. This is different from the viewpoint of Gutzwiller which involves periodic trajectories. The number of our randomly chosen trajectories is 100 or 200. In averaging the Lyapunov exponents, those trajectories with larger Lyapunov exponents will play more significant role. For this, we will consider the work involving one-electronic motion in a lattice and molecular vibration like H_2O.

In this chapter, we will also consider the quantization of DCN vibration by the actions of its periodic trajectories (see Section 3.5). For the two coupled DCN stretching modes, the inter-mode couplings will lead to nonintegrability, i.e., the periodic trajectories will be destroyed accompanying with the birth of new periodic and chaotic trajectories (see Section 1.5 for KAM theorem). Though then, the quantization meets with difficulty, however, it will be pointed out that for not very high excitation, there is a possibility that actions of the remnant periodic trajectories can be employed for quantization, especially of those lower levels. Finally, we explore the Henon-Heiles and AKP systems.

7.2   Hamiltonian for one electron in multiple sites

The Hamiltonian of one electron in multiple sites can be written as:

$$\sum e_i a_i^+ a_i + 1/2 \sum V_{kl} a_k^+ a_l.$$

Here, e_i is the site energy at site i, $a_i^+ a_i$ is the population of electrons or action and $a_k^+ a_l$ is the interaction between sites l and k. $V_{kl}/2$ is its interaction energy. Four-site systems are chosen for study as shown in Fig. 7.1. They are labeled as S1, S2, S3.

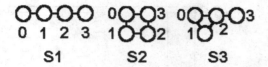

Fig. 7.1   Systems of four-site lattices.

In the coset representation, the system Hamiltonian is

$$
e_0 \left[1 - \sum_{i=1}^{3} (q_i^2 + p_i^2) \right] + \sum_{i=1}^{3} e_i (q_i^2 + p_i^2)
$$

$$
+ \sum_{i=1}^{3} V_{0i} q_i \left[1 - \sum (q_i^2 + p_i^2) \right]^{1/2} + \sum_{i \neq j \neq 0} V_{ij} (q_i q_j + p_i p_j).
$$

Here, e_0, e_1, e_2, e_3 are the site energies at sites 0, 1, 2, 3. e_i corresponds to the α value in HMO (Huckel Molecular Orbital theory). V_{ij} is the interaction energy between sites i and j. $V_{ij}/2$ corresponds to β in HMO. The nearest neighbor approximation will be adopted, i.e., if i, j do not correspond to adjacent sites, then $V_{ij} = 0$. In our calculation, we take $e_i = -8.1 \times 10^4$ cm^{-1}, $V_{ij} = -1.4 \times 10^4$ cm^{-1}. Actions at the sites are: $n_0 = 1 - \sum q_i^2 + p_i^2$ and $n_i = q_i^2 + p_i^2$ $(i = 1, 2, 3)$. Since there is but one electron, we have $\sum q_i^2 + p_i^2 \leq 1$. Angle $\theta_i = \tan^{-1}(-p_i/q_i)$ shows the phase relation between actions n_0 and n_i. θ_i can be correlated with the signs of the linear combination coefficients, C_0 and C_i (The HMO wave function is $\psi = \sum C_i \varphi_i$). If C_0 and C_i are of similar magnitudes and the same sign, then $\theta_i = 0$ (bonding). If they are of opposite signs, then $\theta_i = \pi$ (anti-bonding). If $C_i = 0$, then $\theta_i = \pi/2$ (nonbonding). Furthermore, if $C_i > 0$, (suppose $C_0 > 0$) and its magnitude is quite small, then θ_i is smaller than but very close to $\pi/2$. Similarly, if $C_i < 0$ with quite small magnitude, then θ_i is larger than but very close to $\pi/2$. $\sqrt{n_i}$ is proportional to $|C_i|$.

Let q_i, p_i run over all their allowed ranges, then the maximal and minimal $H(q, p)$ are exactly the highest and lowest level energies of HMO. $H(q, p)$ offers the continuous energy range as shown in Fig. 7.2(a). Our goal is to determine the quantized energies from $H(q, p)$.

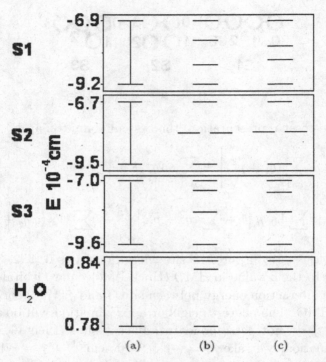

Fig. 7.2   (a) classical energy range (b) quantized energies (c) energies with least $\langle \lambda \rangle$ for S1, S2, S3 and H_2O.

7.3   Quantization: The least averaged Lyapunov exponent

We partition $H(q,p)$ into 100 intervals. Corresponding to each energy at the center of an interval, there are many $(q_1, p_1, q_2, p_2, q_3, p_3)$ points. Thereof, we randomly choose 200 points and calculate their (largest) Lyapunov exponents. These exponents will span a range. Denote their average as $\langle \lambda \rangle$ and consider it as the average of that interval. (If we increase the partitions, the interval will be narrower.) Figure 7.3 shows the relation between $\langle \lambda \rangle$ and energy. Thereof, we have four local minima for S1 and S3. For S2, there are only three local minima. The locations of these minima are shown in Fig. 7.2(c). The energies evidenced by local minimal $\langle \lambda \rangle$ are very close to the quantized energies by HMO (Fig. 7.2(b)). For S2, due to symmetry,

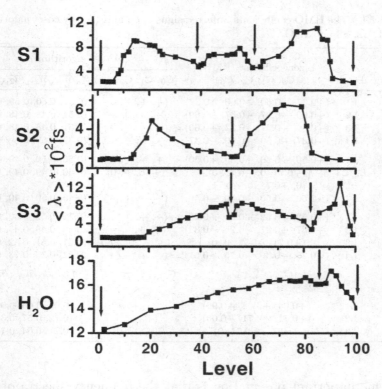

Fig. 7.3   Local minima of averaged Lyapunov exponent $\langle\lambda\rangle$ for S1, S2, S3 and H_2O.

two quantized levels are degenerate. Indeed, in the three local minima of $\langle\lambda\rangle$, there is one that corresponds to the two degenerate levels.

Hence, our conjecture is that quantized energy corresponds to the local minimum of averaged Lyapunov exponent.

This idea shares common ground with those by Bohr and Gutzwiller. In their ideas, only standing wave and stable periodic trajectory can exist in a quantum system. Classically, a stable standing wave and periodic trajectory possess zero Lyapunov exponent. However, here the algorithm does not depend on periodic trajectories. Instead, it depends on the chaotic trajectories possessing larger Lyapunov exponents since they contribute more to $\langle\lambda\rangle$. Hence, the quantization condition proposed is: least chaos in the global phase space.

Table 7.1 The HMO coefficients, inferred signs of C_i and $\sqrt{n_i}$ by coset algorithm for S1, S2, S3 and H_2O.

System	Level	Eigencoefficient (HMO) (C_0, C_1, C_2, C_3)	Coset algorithm (C_0, C_1, C_2, C_3)	$(\sqrt{n_0}, \sqrt{n_1}, \sqrt{n_2}, \sqrt{n_3})$
S1	L1	$(+0.37, +0.60, +0.60, +0.37)$	$(+, +, +, +)$	$(0.37, 0.60, 0.59, 0.37)$
	L2	$(+0.60, +0.37, -0.37, -0.60)$	$(+, +, 0, -)$	$(0.51, 0.48, 0.48, 0.51)$
	L3	$(+0.60, -0.37, -0.37, +0.60)$	$(+, -, -, +)$	$(0.51, 0.48, 0.48, 0.51)$
	L4	$(+0.37, -0.60, +0.60, -0.37)$	$(+, -, +, -)$	$(0.37, 0.59, 0.59, 0.37)$
S2	L1	$(+0.50, +0.50, +0.50, +0.50)$	$(+, +, +, +)$	$(0.50, 0.50, 0.49, 0.50)$
	L2/L3	$(+0.71, 0, -0.71, 0)$ $(0, +0.71, 0, -0.71)$	$(+, 0, -, 0)$	$(0.50, 0.49, 0.49, 0.49)$
	L4	$(+0.50, -0.50, +0.50, -0.50)$	$(+, -, +, -)$	$(0.49, 0.50, 0.49, 0.50)$
S3	L1	$(+0.52, +0.52, +0.61, +0.28)$	$(+, +, +, +)$	$(0.52, 0.52, 0.60, 0.28)$
	L2	$(+0.37, +0.37, -0.25, -0.82)$	$(+, +, 0, -)$	$(0.49, 0.48, 0.47, 0.53)$
	L3	$(+0.71, -0.71, 0, 0)$	$(+, -, -, \sim 0)$	$(0.51, 0.51, 0.52, 0.44)$
	L4	$(+0.30, +0.30, -0.75, +0.51)$	$(+, +, -, +)$	$(0.32, 0.32, 0.73, 0.50)$
H_2O		$(C_{n_s}, C_{n_t}, C_{n_b})$	$(C_{n_s}, C_{n_t}, C_{n_b})$ sign	$(\sqrt{n_s}, \sqrt{n_t}, \sqrt{n_b})$
	L1	$(+0.04, +0.04, +0.99)$	$(+, +, +)$	$(0.08, 0.08, 0.99)$
	L2	$(+0.71, +0.71, -0.06)$	$(+, +, 0)$	$(0.72, 0.57, 0.35)$
	L3	$(+0.71, -0.71, 0)$	$(+, -, 0)$	$(0.72, 0.64, 0.17)$

One important observation is that the frequency spectra of the trajectories (that is the Fourier transform) of the levels with least $\langle \lambda \rangle$ always show simpler patterns than the others.

Once the level energies with least $\langle \lambda \rangle$ are determined, the corresponding $(q_1, p_1, q_2, p_2, q_3, p_3)$ and $(\theta_1, \theta_2, \theta_3)$, (n_0, n_1, n_2, n_3) can be obtained. Numerically, they have a distribution and the maximal values of θ_i and n_i are what we will take to infer the signs of C_i and $\sqrt{n_i}$. These results are listed in Table 7.1. The inferred signs of C_i are the same as those of HMO. Indeed, as C_i of HMO is smaller, the corresponding θ_i is closer to $\pi/2$ and the inferred C_i in the Table is 0. In the Table, though $\sqrt{n_i}$ and C_i of HMO are not exactly identical, their relative magnitudes are consistent. For the 3rd level of S3, the inferred sign of C_2 is negative and its magnitude inferred is the largest, however, it is 0 by HMO. Another inconsistency is for this level $\sqrt{n_3} = 0.44$, while by HMO, C_3 is 0 though C_3 is also assigned as 0 due to $\theta_3 \sim \pi/2$.

7.4   Quantization of H_2O vibration

The algebraic vibrational Hamiltonian of H_2O, $H(q_t, p_t, q_b, p_b)$, can be written as:

$$\omega_s(n_s + n_t + 1) + \omega_b\left(n_b + \frac{1}{2}\right)$$

$$+ X_{ss}\left[\left(n_s + \frac{1}{2}\right)^2 + \left(n_t + \frac{1}{2}\right)^2\right] + X_{bb}\left(n_b + \frac{1}{2}\right)^2$$

$$+ X_{st}\left(n_s + \frac{1}{2}\right)\left(n_t + \frac{1}{2}\right) + X_{sb}(n_s + n_t + 1)\left(n_b + \frac{1}{2}\right)$$

$$+ K_{st}(2n_s)^{\frac{1}{2}} q_t$$

$$+ K_{sb}\{\sqrt{n_s}(q_b^2 - p_b^2) + [q_t(q_b^2 - p_b^2) + 2p_t q_b p_b]/\sqrt{2}\}$$

with

$$n_t = (q_t^2 + p_t^2)/2, \quad n_b = (q_b^2 + p_b^2)/2, \quad n_s = P - n_t - n_b/2.$$

Here, s and t stand for the two O–H stretchings and b for the HOH bending. ω and X are harmonic and anharmonic coefficients. K_{st}, K_{sb} are the coefficients for the 1:1 coupling between s and t and Fermi resonance among s, t and b, respectively. Their elucidation and values have been stated in detail in Section 5.4.

As an example, take $P = 1$. Figure 7.2 shows the energy range by $H(q_t, p_t, q_b, p_b)$ and the three quantum levels by the diagonalization of the Hamiltonian matrix based on the second quantization operators (See Section 2.1.). We partitioned the classical energy range into 100 intervals and randomly chose 200 initial points in the solution space corresponding to the energy at the center of each interval to calculate the Lyapunov exponents and their average $\langle \lambda \rangle$. The results are shown in Fig. 7.3. Indeed, there are three local minima. These level energies by locally minimal Lyapunov exponents are also shown in Fig. 7.2(c). They are very close to the quantum energies (Fig. 7.2(b))

From the energies corresponding to minimal $\langle \lambda \rangle$, we can obtain (q_t, p_t, q_b, p_b) and θ_i, n_i from which the signs of C_i and magnitudes $\sqrt{n_i}$ can be obtained. As shown in Table 7.1, the results are quite consistent with those by quantum mechanics.

7.5   Action integrals of periodic trajectories: The DCN case

In Section 3.5.2, we have analyzed the periodic trajectories for the coupled D–C and C–N nonlinear oscillators of DCN. For periodic trajectories, we can employ two Poincare surfaces of section for analysis: (q_s, p_s), $q_t = 0$, $p_t < 0$ and (q_t, p_t), $q_s = 0$, $p_s < 0$ and define the number of points left on the surface as their periods.

For convenience, we will use p^n or q^m to show that a periodic trajectory has n and m points on the surfaces of (q_s, p_s) and (q_t, p_t), respectively. For a periodic trajectory, there are two notions p^n and q^m. m is different from n, in general except for period-1 trajectories. For example, for p^3, (period-3 trajectory) we have q^4. For p^5, p^7, p^8, p^9, p^{12}, p^{15}, p^{18}, we have q^7, q^9, q^{11}, q^{12}, q^{16}, q^{20} and q^{24} (for p^5, we have also q^5). m and n show a nice linear relation as shown in Fig. 7.4. The slope is 1.313. This value is very close to the harmonic frequency ratio of D–C and C–N: 1.3. In fact, this is the so-called winding number.

For an eigenstate, we may have several trajectories possessing the same $p^n(q^m)$. For period-1 trajectories, we have two different p^1 (q^1) trajectories. Their actions are concentrated on either D–C or

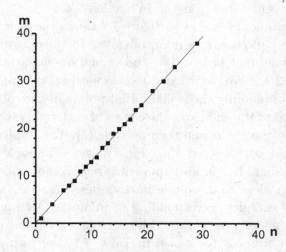

Fig. 7.4   The relation between the periods n and m on the surfaces: $(q_s\ p_s)$ and (q_t, p_t) for periodic trajectories.

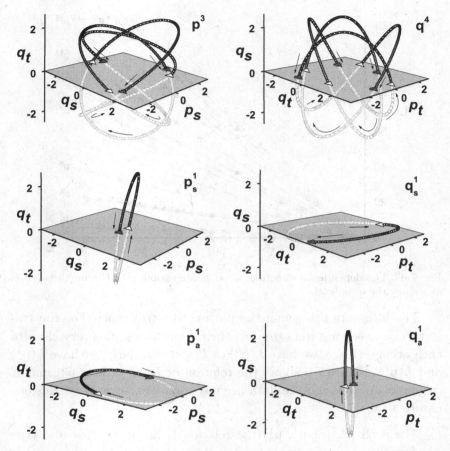

Fig. 7.5 The period-1 (L11) and period-3 (L22) trajectories of DCN and their intersections on the (q_s, p_s) and (q_t, p_t) surfaces. ▲ denotes p_s or $p_t < 0$. △ denotes p_s or $p_t > 0$. Arrows show the direction of the evolving trajectories.

C–N and their displacements are, respectively, antisymmetric and symmetric. Hence, we will use p_a^1 (q_a^1) and p_s^1 (q_s^1) to represent them. Figure 7.5 shows period-1 (L11) and period-3 (L22) trajectories and their intersections on the (q_s, p_s) and (q_t, p_t) surfaces. (L# shows the level numbering.)

For a periodic trajectory, its action integral, L, is defined as

$$L = \frac{1}{2\pi} \oint \vec{p} \cdot d\vec{q} = \frac{1}{2\pi} \left(\oint p_s dq_s + \oint p_t dq_t \right).$$

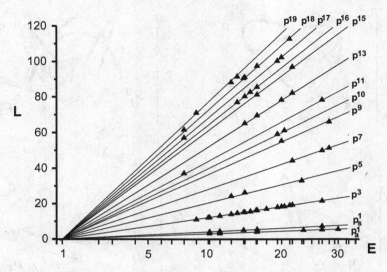

Fig. 7.6   The dependence of action, L on E for various p^n. The numbers on E axis show the eigenlevels.

The integration is along the path of the trajectory. For the trajectories possessing the same p^n, their L are the same (very close to each other). Hence, we have $L(p^n)$ or $L(q^m)$. Similarly, we have $L(p_s^1)$ and $L(p_a^1)$. Figure 7.6 shows the relation of L (of various p^n) and E (eigen-level). Though the data are not complete, several conclusions can be drawn:

(1) For each p^n, (or p_s^1, p_a^1) the relation is linear. This is also true for q^m.

(2) For each E, L is linear against n and m. Figure 7.7 shows the situation for the 17th level (levels are counted from the lowest one). We also note that $L(p_s^1)$ and $L(p^n)$, $L(p_a^1)$ and $L(q^m)$ show a linear relation against period. Specifically, we have:

$$L(p^i) : L(p^j) \cong i : j$$

$$L(q^s) : L(q^t) \cong s : t$$

(3) For lower quantum states and smaller n, $L(p^n)$ $(L(q^m))$ is, or close to, an integer.

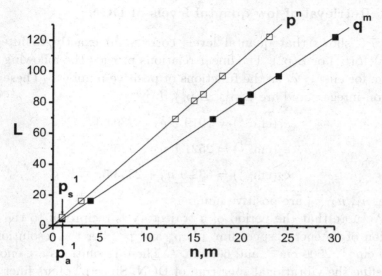

Fig. 7.7 The dependence of $L(p^n)$, $L(q^m)$ on periods n and m for the 17th level of DCN.

(4) From Fig. 7.6, we have for the ground state:

$$L(p^n) = L(p_s^1) = L(p_a^1) = 0.$$

The phase space of the ground state is only a point in which all the trajectories shrink to a point.

These rich linear behaviors enable us to calculate all $L(p^n)$ at any E from very few L values. For instance, suppose $L(p_s^1)$, $L(p_a^1)$ and $L(p^3)$ at any two E_1, E_2 are known, then from Fig. 7.6, $L(p_s^1)$, $L(p_a^1)$ and $L(p^3)$ at any E can be known. Meanwhile, for any E, since $L(p_s^1)$ and $L(p^3)$ are known, all $L(p^n)$ will be known just like the situation shown in Fig. 7.7. Furthermore, from $L(p_a^1)$ and $L(q^4)$ ($= L(p^3)$), all $L(q^m)$ will be known. (These $L(q^m)$ and $L(p^n)$ are interrelated.)

In the next section, we will show that these relations give us a simple and easy way to retrieve the low quantal levels.

7.6   Retrieval of low quantal levels of DCN

Fig. 7.6 shows that quantal levels correspond exactly to integral $L(p_s^1)$, $L(p_a^1)$ or $L(p^3)$. The linear relations provide the following formulae for energy e_i as the functions of positive numbers (These can be non-integers and are $L(p_s^1)$, $L(p_a^1)$, $L(p^3)$):

$$e_1(\text{cm}^{-1}) = 1919.1\, n_1 + 2304.4$$

$$e_2(\text{cm}^{-1}) = 2521.7\, n_2 + 2453.4$$

$$e_3(\text{cm}^{-1}) = 632.9\, n_3 + 2343.5.$$

Here, n_1, n_2, n_3 are positive numbers.

We note that the period of a trajectory is reciprocal to the resolution of its energy spectrum. p_s^1, p_a^1 and p^3 offer the resolution of 1900 cm^{-1}, 2500 cm^{-1} and 600 cm^{-1}. These resolutions are enough to define the vibrational spectrum of DCN. Since p^3 offers finer resolution than by p_s^1 and p_a^1, n_3 will be defined by n_1 and n_2, i.e., the levels defined by n_3 should be in between the levels by n_1 and n_2.

Quantal levels can be retrieved by the following procedures:

(i) Let $n_1 = 0, 1, 2, \ldots$, we have the levels as shown in Fig. 7.8(a).

(ii) Let $n_2 = 1, 2, \ldots$, we have the levels as shown in Fig. 7.8(b). (Note $n_2 = 0$ and $n_1 = 0$ offer quite the same level)

(iii) Among the levels in Figs. 7.8 (a) and (b) that are connected by the broken lines (corresponding to $n_1 = n_2 \equiv n_0$), insert proper e_3 levels (corresponding to $n_3 = 3n_0 + 1, 3n_0 + 2, \ldots, 4n_0 - 1$) as shown in Fig. 7.8 (c). Fig. 7.8 (d) is the totality of the levels in Figs. 7.8 (a), (b) and (c). They are very close to the quantal levels shown in Fig. 7.8 (e). Table 7.2 shows these level energies. The deviation is not larger than 100 cm^{-1}, i.e., less than 1%. We have to stress that this method is suitable to low levels, less than 15000 cm^{-1} since there is difficulty in finding the periodic trajectories for higher levels.

In summary, the whole procedure starts from two randomly chosen energies which are very probably not the eigenenergies. Determine their p^3, p_s^1 and p_a^1 trajectories and the actions. Follow the method shown in the last section to determine the linear dependence

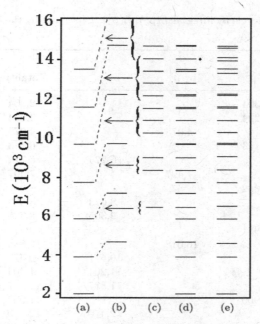

Fig. 7.8　The quantal levels by integral (a) $L(p_s^1)$, (b) $L(p_a^1)$, (c) $L(p^3)$. Arrows and brackets show that in between the levels in (a), (b) connected by the broken lines, we can insert the levels with integral $L(p^3)$. (d) is the totality of the levels in (a), (b) and (c). In (e) are the quantal levels. '•' denotes the degenerate levels.

of $L(p_s^1)$, $L(p_a^1)$ and $L(p^3)$ on energy. Then, from the linear functions, quantized energies can be elucidated.

For DCN, 1:1 and 2:3 couplings between D–C and C–N stretchings destroy the otherwise conserved quantum numbers such as $n_s + n_t$, $n_s/2 + n_t/3$. However, the destruction is not so serious that they are still approximately conserved. (See Chapter 5). An interesting point is that $L(p_s^1)$ and $L(p_a^1)$ of the levels in Figs. 7.8(a) and (b) are just the approximate constants of motion: $n_s + n_t$ (They are the polyad number P_1 in Section 5.6). This implies the close relation between $L(p^n)$ and the (approximately) conserved quantum number. Their exact relation can bear important significance for the systems of highly excited vibration.

For a nonintegrable system, due to the destruction of periodic trajectories, its quantization via action integrals was suspected. Einstein

Table 7.2 The levels by integral $L(p_s^1)$, $L(p_a^1)$, $L(p^3)$ and by the quantal method (cm^{-1}).

Level	by p_s^1	by p_a^1	by p^3	Totality	Quantal method
1	2,304			2,304	2,301
2	4,224			4,224	4,221
3		4,975		4,975	4,934
4	6,143			6,143	6,139
5			6,774	6,774	6,821
6		7,497		7,497	7,521
7	8,062			8,062	8,060
8			8,673	8,673	8,705
9			9,305	9,305	9,364
10	9,981			9,981	9,984
11		10,018		10,018	10,062
12			10,571	10,571	10,599
13			11,204	11,204	11,200
14			11,837	11,837	11,852
15	11,900			11,900	11,915
16			12,470	12,470	12,502
17		12,540		12,540	12,556
18			13,103	13,103	13,051
19			13,736	13,736	13,628
20	13,819			13,819	13,851
21			14,369	14,369	14,278
22			14,369	14,369	14,425
23			15,002	15,002	14,915
24		15,062		15,062	15,001

was the first to point out this difficulty for a nonintegrable system. Our work shows that for low levels, the situation may not be so pessimistic. The remnant periodic trajectories in a chaotic system possess quantities that are related to quantization. In this aspect, Gutzwiller's idea about periodic trajectories is crucial. On the other hand, we have the conjecture that quantization can be based on the locally minimal Lyapunov exponents. That is: quantization is related to the minimal degree of chaos in a global sense. The implication is that chaotic and periodic trajectories are intrinsically related to each other in a nonintegrable system. Though the KAM theorem demonstrates the evolution of trajectories under perturbation, there is still much that we do not know and awaits future exploration.

7.7 Quantization of Henon-Heiles system

In this section, the periodic trajectories and their action integrals of the Henon-Heiles system of two coupled oscillators will be studied. The results show that there is linear relationship between the action integral and the system energy. Hence, we can semi-classically quantize the system by the integral action integrals. This demonstrates that the remnant periodic trajectories bear the properties that are related to the quantization of a chaotic system. This helps our understanding of the classical properties embedded in a quantum system.

Henon-Heiles system is a typical three-body model in astrophysics[2] possessing a symmetric potential. Its Hamiltonian is

$$H = \frac{1}{2}(p_x^2 + p_y^2 + x^2 + y^2) + \lambda x \left(y^2 - \frac{1}{3}x^2\right).$$

Here, x and y are the coordinates. p_x and p_y are the conjugate momenta. λ is the coupling constant. By transforming to the polar coordinates via $x = r\cos\theta$ and $y = r\sin\theta$, the Hamiltonian becomes:

$$H = \frac{1}{2}(p_r^2 + r^2 + p_\theta^2/r^2) - \frac{1}{3}\lambda r^3 \cos 3\theta.$$

For $H(x, p_x, y, p_y)$, the equations of motion are:

$$\frac{d}{dt}\begin{pmatrix} x \\ y \\ p_x \\ p_y \end{pmatrix} = \begin{pmatrix} p_x \\ p_y \\ -x - \lambda(y^2 + x^2) \\ -y - 2\lambda xy \end{pmatrix}.$$

Obviously, its potential is of 120^0 rotational symmetry. In the following calculation, we will take $\lambda = 0.1118$.

We will use Poincare surface of section (x, p_x) $(y = 0, p_y < 0)$ to view its dynamics. We found that as the system energy is $0 < E < 11.972$, the dynamics is very rich. As E is small, the system is integrable and the phase space is full of periodic trajectories. As the

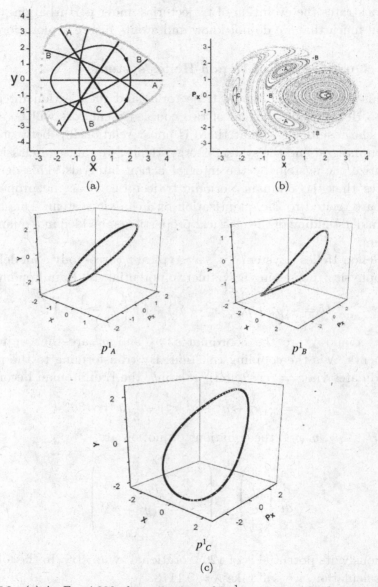

(a)   (b)

p_A^1 p_B^1

p_C^1

(c)

Fig. 7.9   (a) As $E = 4.986$, the projections of 8 p^1 trajectories on the x-y surface. The 2 degenerate p_C^1 are clockwise and counter-clockwise. The outer potential curve is of energy $V = 4.986$. (b) As $E = 4.986$, the traces of p_A^1, p_B^1 and p_C^1 on Poincare surface of section (x, p_x) $(y = 0, p_y < 0$. They are a point. (c) The 3 dimensional projections of p_A^1, p_B^1, p_C^1 as $E = 4.986$. (d) The x-y and 3 dimensional (x, y, p_x) projections of p_S^1 with $E = 14.0$.

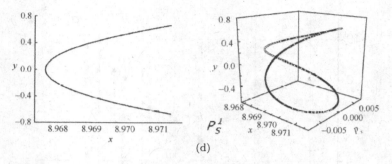

Fig. 7.9 (*Continued*)

system energy increases, more periodic trajectories are destroyed and the chaotic regions start appearing. Finally, as $E = 11.972$, the phase space is mostly chaotic. Then there are two small quasi-periodic islands immersed in the chaotic sea.

Our goal is to quantize the system via periodic trajectories. The iterative segment intersection method[3] was employed to find out the periodic trajectories. For convenience, p^1 is used for period 1 trajectory. Four kinds of p^1 were found and named as: A("apex"), B("base"), C("circle") which stay in the bottom of the potential basin and S("saddle") which appears around the saddle point. A, B and S are liberation while C is precession. Since the potential possesses three-fold rotational symmetry, totally, there are 3 A's, 3 B's, 2 C's and 3 S's trajectories. These p^1 trajectories are denoted as p_A^1, p_B^1, p_C^1, p_S^1, respectively.

The potential at the saddle point is 11.972. Only above this potential, there will be p_S^1. As E < 11.972, there are 8 p_A^1, p_B^1 and p_C^1 as shown in Fig. 7.9(a) where there are 3 degenerate p_B^1 and p_A^1, and 2 degenerate p_C^1 (clockwise and counter-clockwise). Fig. 7.9(b) shows their traces on the Poincare surface of section, (x, p_x) $(y = 0, p_y < 0)$. Fig. 7.9(c) shows their projections on the 3 dimensional phase space. Fig. 7.9(d) shows the motion of p_S^1 near the saddle point.

The action integral, L, of a periodic trajectory is defined as

$$L = \frac{1}{2\pi} \oint \vec{p}_\alpha \cdot d\alpha = \frac{1}{2\pi} \left(\oint p_x dx + \oint p_y dy \right).$$

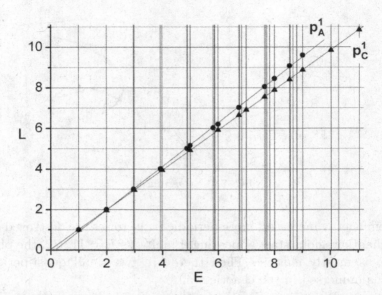

Fig. 7.10   The action integrals, $L(p_A^1)$, $L(p_C^1)$ as the function of energy E. The vertical lines show where the eigen-levels are. For clarity, $L(p_B^1)$ is not plotted in.

We note that for the eigenenergy, $L(p_A^1)$, $L(p_B^1)$, $L(p_C^1)$, $L(p_S^1)$ are very close to integers, as shown in Fig. 7.10.

We can now infer the eigenenergies by the integral action integrals. The relations are:

$$e_1 = 0.932149 \, n_1 + 0.1474846$$

$$e_2 = 1.0186 \, n_2 - 0.0296$$

$$e_3 = 0.9513 \, n_3 + 0.0905$$

where n_1, n_2, n_3 are positive integers (They are, in fact, $L(p_A^1)$, $L(p_C^1)$, $L(p_B^1)$) and e_i is the corresponding energy. These relations are obtained by the fit to the data in Fig. 7.10. It is found that to an eigenenergy, there must be the corresponding n_1, n_2 or n_3. However, the inverse does not hold. n_3 has to be defined by n_1 and n_2. This is because that p_B^1 offers finer spectral resolution than those by p_A^1 and p_C^1, i.e., the eigenenergies offered by n_3 have to be in between those offered by n_1 and n_2. In conclusion, the following steps offer

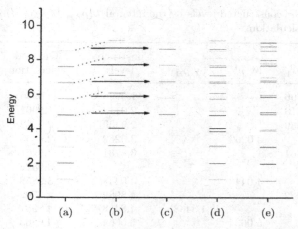

Fig. 7.11   (a) the levels constructed by integral $L(p_A^1)$, (b) by $L(p_C^1)$, (c) by $L(p_B^1)$. Arrows show that in between the levels in (a) and (b) connected by the broken lines, proper levels by the integral $L(p_B^1)$ are inserted. (d) is the totality of the levels in (a), (b), (c). (e) is by quantal calculation.

the construction of the eigenenergies:

(i) let $n_1 = 0, 1, 2, \ldots$, the levels shown in 7.11(a) are constructed.

(ii) let $n_2 = 3, 4, \ldots$, the levels shown in 7.11(b) are constructed.

(iii) In between the levels connected by the broken lines in Figs. 7.11(a) and (b), insert the proper e_3 levels (corresponding to $n_3 = 5, 6, 7 \ldots$,) as shown in Fig. 7.11(c) by arrows.

Fig. 7.11(d) shows the totality of the levels in Fig. 7.11(a),(b),(c). These levels are very close to those by quantal calculation. Table 7.3 shows that except the lowest level, the errors are less than 2 %. For higher levels, since the linear relation between action integral and energy no longer holds, this algorithm does not work.

The roles by p_A^1 and p_C^1 are similar to p_s^1 and p_a^1 in the DCN case (Section 7.6). p_A^1 and p_C^1 offers the upper and lower bounds: levels 1, 2, 3, 5, 8, 11, 15, 20, 25 (Fig. 7.11(a)) and 4, 7, 10, 14, 19, 24, 30 (Fig. 7.11(b)), respectively. Thus the spectrum can be grouped into 9 sets.

Table 7.4 shows the semi-classical quantal levels by D. W. Noid and R. A. Marcus.[4] These levels are grouped by the principal

Table 7.3   The constructed levels by the integral $L(p_A^1)$, $L(p_C^1)$, $L(p_B^1)$ and those by quantal calculation.

Level	by p_A^1	by p_C^1	by p_B^1	Constructed levels	Quantal calculation	Deviation
1	1.0796			1.0796	0.9986	0.08111
2	2.0118			2.0118	1.9901	0.0109
3	2.9439			2.9439	2.9562	−0.00416
4		3.0262		3.0262	2.9853	0.0137
5	3.8761			3.8761	3.926	−0.01271
6					3.9824	
7		4.0448		4.0448	3.9858	0.0148
8	4.8082			4.8082	4.8702	−0.01273
9			4.8468	4.8468	4.8987	−0.01059
10		5.0634		5.0634	4.9863	0.01546
11	5.7404			5.7404	5.817	−0.01317
12			5.798	5.798	5.867	−0.01176
13					5.8815	
14		6.082		6.082	5.9913	0.01514
15	6.6725			6.6725	6.7379	−0.00971
16			6.7493	6.7493	6.7649	−0.00231
17					6.8354	
18					6.9989	
19		7.1006		7.1006	6.9994	0.01446
20	7.6047			7.6047	7.6595	−0.00715
21					7.6977	
22			7.7006	7.7006	7.7369	−0.00469
23					7.8327	
24		8.1192		8.1192	8.0094	0.01371
25	8.5368			8.5368	8.5541	−0.00202
26					8.5764	
27			8.6518	8.6518	8.6779	−0.00301
28					8.8113	
29					8.8152	
30		9.1377		9.1377	9.0217	0.01286

quantum number n with different angular quantum numbers l ($l = 0$, $\pm 1, \ldots \pm n$). What are the roles played by p_A^1 and p_C^1? As seen from Fig. 7.9(a), p_C^1 possesses the least radial momentum p_r (This can be confirmed by using the polar coordinates.) while p_A^1 possesses the least angular momentum p_θ. Hence, we can employ p_A^1 to reconstruct the levels sharing a common n with the least angular momentum. Similarly, the levels of the largest angular momentum

Table 7.4 The semi-classical quantal levels of Henon-Heiles system by D. W. Noid and R. A. Marcus.[4]

Level	(l, n)	Scaled energy
1	(0, 0)}	0.9986
2	(±1, 1)}	1.9901
3	(0, 2) ⎫	2.9562
4	(±2, 2) ⎭	2.9853
5	(±1, 3) ⎫	3.9260
6	(±3, 3) ⎭	3.9824
7	⋯	3.9858
8	(0, 4) ⎫	4.8702
9	(±2, 4) ⎬	4.8987
10	(±4, 4) ⎭	4.9863
11	(±1, 5) ⎫	5.8170
12	(±3, 5) ⎪	5.8670
13	⋯ ⎬	5.8815
14	(±5, 5) ⎭	5.9913
15	(±0, 6) ⎫	6.7379
16	(±2, 6) ⎪	6.7649
17	(±4, 6) ⎬	6.8354
18	(±6, 6) ⎭	6.9989
19	⋯	6.9994
20	(±1, 7) ⎫	7.6595
21	(±3, 7) ⎪	7.6977
22	⋯ ⎬	7.7369
23	(±5, 7) ⎪	7.8327
24	(±7, 7) ⎭	8.0094
25	(0, 8) ⎫	8.5541
26	(±2, 8) ⎪	8.5764
27	(±4, 8) ⎪	8.6779
28	(±6, 8) ⎬	8.8113
29	⋯ ⎪	8.8152
30	(±8, 8) ⎭	9.0217
31	(±9, 9)}	10.0354
32	⋯	10.0356
33	(0, 10) ⎫	10.3052
34	(±10, 10) ⎭	11.0497

can be reconstructed by p_C^1. We note that p_r and p_θ of p_B^1 are in between those of p_A^1 and p_C^1. Hence, via p_B^1, the levels in between those by p_A^1 and p_C^1 can be reconstructed is not unexpected. This is what we have done as shown in Fig. 7.11.

It seems that p_S^1 does not play the role in reconstructing the levels. For the moment, only 4 kinds of periodic trajectories are identified. However, we expect that more periodic trajectories not yet known can be found and may play the roles in reconstructing the levels.

7.8 Quantal correspondence in the classical AKP system

Anisotropic Kepler Problem (AKP) describes the system of an electron in an anisotropic Coulomb field. This system is simple but possesses emerging chaos. The system Hamiltonian is

$$H = \frac{P_\rho^2}{2} + \gamma \frac{P_z^2}{2} - \frac{e^2}{\sqrt{\rho^2 + z^2}}.$$

The parameter γ differentiates the anisotropy along x-y plane and z axis. $\gamma = 1$ corresponds to the case of isotropy. Then, the system is integrable. Otherwise, it is non-integrable. We make the coordinate transformation[5] to avoid the singularity at $\rho = z = 0$:

$$\rho = z_1 z_2; \qquad p_\rho = \frac{1}{z_1^2 + z_2^2}(z_1 z_4 + z_2 z_3)$$

$$z = \frac{1}{2}(z_2^2 - z_1^2); \qquad p_z = \frac{1}{z_1^2 + z_2^2}(z_2 z_4 - z_1 z_3).$$

The transformed Hamiltonian in (z_1, z_2, z_3, z_4) is:

$$H' = 2 = \frac{1}{2}(z_3^2 + z_4^2) - \varepsilon(z_1^2 + z_2^2) + (\gamma - 1)\frac{(z_1 z_3 - z_2 z_4)^2}{2(z_1^2 + z_2^2)}$$

ε is the system energy. As $\gamma = 1$,

$$\varepsilon = \frac{1}{2n^2}, \quad n = 1, 2, 3, \ldots.$$

The equations of motion are

$$\frac{d}{dt}\begin{pmatrix} z_1 \\ z_2 \\ z_3 \\ z_4 \end{pmatrix} = \begin{pmatrix} z_3 + (\gamma - 1)z_1\dfrac{(z_1 z_3 - z_2 z_4)}{(z_1^2 + z_2^2)} \\ z_4 - (\gamma - 1)z_2\dfrac{(z_1 z_3 - z_2 z_4)}{(z_1^2 + z_2^2)} \\ 2\varepsilon z_1 - (\gamma - 1)\dfrac{(z_3^2 z_1 z_2^2 + z_2 z_3 z_4(z_1^2 - z_2^2) - z_2^2 z_1 z_4^2)}{(z_1^2 + z_2^2)} \\ 2\varepsilon z_2 - (\gamma - 1)\dfrac{(z_4^2 z_2 z_1^2 - z_1 z_3 z_4(z_1^2 - z_2^2) - z_1^2 z_2 z_3^2)}{(z_1^2 + z_2^2)} \end{pmatrix}.$$

Shown in Fig. 7.12 is the structure of Poincare surface of section as $\gamma = 0.95$, $\varepsilon = -0.5$. The period 1 trajectories can be identified in which B and C are stable and A is unstable. Note that C is in (z_2, z_4) since its (z_1, z_3) are 0 always.

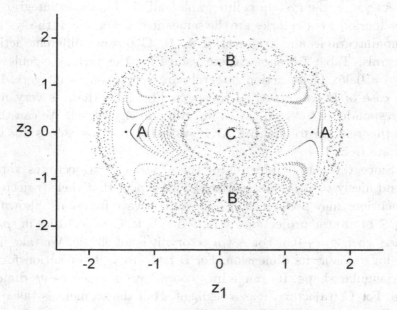

Fig. 7.12 The structure of Poincare surface of section (z_1, z_3) $(z_2 = 0, z_4 < 0)$ as $\gamma = 0.95$, $\varepsilon = -0.5$. A, B, C are periodic trajectories of period 1.

Table 7.5   The action integrals L(A), L(B), L(C) for the various energies and the dimensions of the periodic trajectories, R as $\gamma = 0.95$. n is positive integer.

ε	L(A)	L(B)	L(C)	n^2	R(A)	R(B)	R(C)
−0.5	1.01	1.00	1.03	1	1.04	1.02	1.00
−0.125	2.03	2.01	2.05	4	4.07	4.00	4.00
−0.05556	3.05	3.01	3.08	9	9.19	8.98	9.00
−0.03125	4.07	4.02	4.10	16	16.17	15.98	16.00
−0.02	5.04	5.02	5.13	25	25.39	24.97	25.00
−0.01389	6.12	6.02	6.16	36	36.61	35.94	36.00
−0.0102	7.14	7.03	7.18	49	49.93	48.94	49.02

The action integral is defined as:

$$L = \frac{1}{2\pi} \oint \vec{p} \cdot d\vec{q} = \frac{1}{2\pi} \left(\oint p_\rho d\rho + \oint p_z dz \right)$$

$$= \frac{1}{2\pi} \left(\oint z_3 dz_1 + \oint z_4 dz_2 \right).$$

As $\gamma = 1$, the system is integrable and all the action integrals of these period 1 trajectories are the same integer. For $\gamma \neq 1$, the system is nonintegrable and trajectories A, B, C possess different action integrals. Table 7.5 shows these values for the various eigenlevels as $\gamma = 0.95$. These action integrals are very close to integers. For the case of B trajectory, as shown in Fig. 7.13, there is very nice correspondence between action integral and eigen level. We can thus use this relation to quantize AKP system at least as γ does not too deviate from 1.

Since the action integrals of these period 1 trajectories correspond nicely to the quantized levels, we expect that their trajectory dimensions may possess connection to the wave functions. Shown in Fig. 7.14 are the projections of period 1 A, B, C trajectories in (ρ, z) space when $\varepsilon = -0.5$. For A trajectory, it is an ellipse. We take half the long axis as its dimension. For B trajectory, its evolution forms a triangular shape. Its range in ρ coordinate is taken as its dimension. For C trajectory, it is a segment. Half the segment is taken as its dimension. Table 7.5 shows these dimensions, R for the levels. Obviously, we have the relation $R = n^2$, with $n = 1, 2, 3 \ldots$. This is

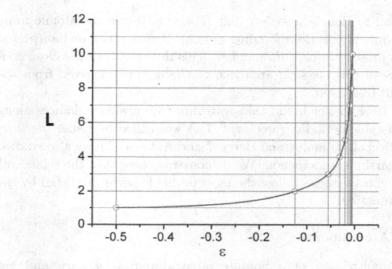

Fig. 7.13 Action integral of B trajectory as a function of energy ε. The vertical lines show the locations of the eigen levels.

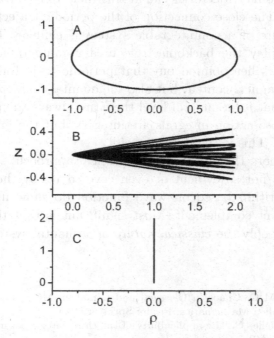

Fig. 7.14 The projections of period 1 A, B, C trajectories in (ρ, z) space as $\varepsilon = -0.5$, $\gamma = 0.95$.

identical to Bohr's assertion that $R = n^2 a_0$ (Since the atomic units are adopted here, Bohr's radius $a_0 = 1$). Hence, n can be interpreted as the principal quantum number. This demonstrates the close connection of this classical approach to the quantity derived from the quantal treatment.

In 1984, Heller found that wave function always enhances along the unstable periodic trajectory.[6] This was called by Heller the scarring effect. Bogomolny and Berry[7,8] and Antonsen[9] have also studied this scarring phenomenon. We demonstrate here that the (classical) periodic trajectories follow the most probable paths predicted by the wave function.

7.9 A comment

The quantization of a nonintegrable system is a hard and yet unsolved issue. Our work shows that the action integrals of the remnant periodic trajectories can at least help to reconstruct the low lying levels. The close connection of the periodic trajectories to the quantal nature of a nonintegrable system is evident. The periodic trajectories play the backbone role in a nonintegrable or chaotic system. Gutzwiller pointed out that periodic trajectories not only bear the dynamical content but also the quantal characteristics. The problem is that for a quantal level there are always the periodic trajectories whose action integrals are integers, but the inverse hardly holds always. This is the difficult point.

We also note that by this semi-classical approach, some characters that are not so evident or even covered up by the traditional quantal algorithm based on wave function do show up. This is a very important complement. Most significant is that these characters reflect deeply the *classical nature* of a quantal system.

References

1. Gutzwiller M C. Chaos in Classical and Quantum Mechanics in Interdisciplinary Applied Mathematics. Berlin: Springer, 1990.
2. Henon M, Heiles C. The applicability of the third integral of motion. *Astron. J.*, 1964, 69: 449.
3. Gao J, Delos J B. *Phys. Rev. A*, 1994, 49: 869.

4. Noid D W, Marcus R A. *J. Chem. Phys.*, 1977, 67: 559.
5. Ramasubramanian K, Sriram M S. *Physica D*, 2000, 139: 72.
6. Heller E J. *Phys. Rev. Lett.*, 1984, 53: 1515.
7. Bogomolny E B. *Physica D*, 1988, 31: 169.
8. Berry M V. *Proc. Roy. Soc. A*, 1989, 423: 219.
9. Antonsen T M, Ott E, Chen Q, Oerter R. *Phys. Rev. E*, 1995, 51: 111.

Chapter 8

Dynamics of DCO/HCO and Dynamical Barrier Due to Extremely Irrational Couplings

8.1 The coset Hamiltonian of DCO

With the advent of laser technology, the dynamics of highly excited vibration is being under exploration. DCO (deuterated formyl radical) is such a system that has been studied heavily both experimentally and theoretically.[1-5] The 1:2:1 (among D-C and C-O stretching and bending motions) resonance was confirmed. Based on the levels (states) reported in Ref. [2], Troellsch *et al.*[3] employed an algebraic Hamiltonian to study its kinetics of the intramolecular vibrational energy redistribution (IVR). We will treat this issue from a *global* viewpoint. That is, we will explore the dynamics of the whole levels sharing a common polyad number according to the 1:2:1 resonance in a dynamical phase (coset) space. In this way, the classical nonlinear dynamical ideas such as dynamical potential and Lyapunov exponent can be employed for its dynamical analysis.

The algebraic Hamiltonian by Troellsch *et al.* can be cast in the coset space as:

$$H(q_1, p_1, q_3, p_3)$$

$$= \omega_1 \left(\frac{p_1^2 + q_1^2}{2} + \frac{1}{2} \right) + X_{11} \left(\frac{p_1^2 + q_1^2}{2} + \frac{1}{2} \right)^2$$

$$+ \omega_2 \left(2P - p_1^2 - q_1^2 - p_3^2 - q_3^2 + \frac{1}{2} \right)$$

$$+ X_{22} \left(2P - p_1^2 - q_1^2 - p_3^2 - q_3^2 + \frac{1}{2} \right)^2$$

$$+ \omega_3 \left(\frac{p_3^2 + q_3^2}{2} + \frac{1}{2} \right) + X_{33} \left(\frac{p_3^2 + q_3^2}{2} + \frac{1}{2} \right)^2$$

$$+ X_{12} \left(\frac{p_1^2 + q_1^2}{2} + \frac{1}{2} \right) \left(2P - p_1^2 - q_1^2 - p_3^2 - q_3^2 + \frac{1}{2} \right)$$

$$+ X_{13} \left(\frac{p_1^2 + q_1^2}{2} + \frac{1}{2} \right) \left(\frac{p_3^2 + q_3^2}{2} + \frac{1}{2} \right)$$

$$+ X_{23} \left(2P - p_1^2 - q_1^2 - p_3^2 - q_3^2 + \frac{1}{2} \right) \left(\frac{p_3^2 + q_3^2}{2} + \frac{1}{2} \right)$$

$$+ K_{1,0,-1} \left[1 + \lambda_1 \left(\frac{p_1^2 + q_1^2}{2} \right) + \lambda_2 \left(\frac{p_3^2 + q_3^2}{2} \right) \right] (q_1 q_3 + p_1 p_3)$$

$$+ K_{1,-2,0} \sqrt{2} q_1 (2P - p_1^2 - q_1^2 - p_3^2 - q_3^2)$$

$$+ K_{0,2,-1} \sqrt{2} q_3 (2P - p_1^2 - q_1^2 - p_3^2 - q_3^2)$$

$$+ K_{2,0,-2} \left[\frac{1}{2} (q_1^2 - p_1^2)(q_3^2 - p_3^2) + 2 q_1 q_3 p_1 p_3 \right]$$

where $n_2 = 2(P - n_1 - n_3)$, $n_1 = (q_1^2 + p_1^2)/2$, $n_3 = (q_3^2 + p_3^2)/2$. Subscripts 1, 2 and 3 denote, respectively, the variables for D-C stretching, the bending and C-O stretching. The coefficients are determined by fit to the levels in Ref. [2] and are listed in Ref. [3]. They are: $\omega_1 = 2099.4 \, \text{cm}^{-1}$, $\omega_2 = 871.1 \, \text{cm}^{-1}$, $\omega_3 = 1863.6 \, \text{cm}^{-1}$, $X_{11} = -106.3 \, \text{cm}^{-1}$, $X_{12} = -27.6 \, \text{cm}^{-1}$, $X_{13} = 7.4 \, \text{cm}^{-1}$, $X_{22} = -4.7 \, \text{cm}^{-1}$, $X_{23} = -5.2 \, \text{cm}^{-1}$, $X_{33} = -11.5 \, \text{cm}^{-1}$, $K_{101} = 42.25 \, \text{cm}^{-1}$, $K_{120} = 14.50 \, \text{cm}^{-1}$, $K_{021} = -3.96 \, \text{cm}^{-1}$,

$K_{202} = -3.85\,\text{cm}^{-1}$, $\lambda_1 = 0.34$, $\lambda_2 = -0.11$. This algebraic Hamiltonian is based on coupled Morse oscillators with 1:1 and 2:2 resonances between D-C and C-O stretchings and Fermi resonance among the three modes. Notably, the action $P = n_1 + n_2/2 + n_3$ is a conserved quantity under this Hamiltonian. There are $(P+1)(P+2)/2$ levels for each P. We note that we may have $H(q_1, p_1, q_2, p_2)$, alternatively, by changing the coset variables.

$$
\begin{aligned}
&H(q_1, p_1, q_2, p_2) \\
&= \omega_1 \left(\frac{p_1^2 + q_1^2}{2} + \frac{1}{2} \right) + X_{11} \left(\frac{p_1^2 + q_1^2}{2} + \frac{1}{2} \right)^2 \\
&\quad + \omega_2 \left(\frac{p_2^2 + q_2^2}{2} + \frac{1}{2} \right) + X_{22} \left(\frac{p_2^2 + q_2^2}{2} + \frac{1}{2} \right)^2 \\
&\quad + \omega_3 \left(P - \frac{p_1^2 + q_1^2}{2} - \frac{p_2^2 + q_2^2}{4} + \frac{1}{2} \right) \\
&\quad + X_{33} \left(P - \frac{p_1^2 + q_1^2}{2} - \frac{p_2^2 + q_2^2}{4} + \frac{1}{2} \right)^2 \\
&\quad + X_{12} \left(\frac{p_1^2 + q_1^2}{2} + \frac{1}{2} \right) \left(\frac{p_2^2 + q_2^2}{2} + \frac{1}{2} \right) \\
&\quad + X_{13} \left(\frac{p_1^2 + q_1^2}{2} + \frac{1}{2} \right) \left(P - \frac{p_1^2 + q_1^2}{2} - \frac{p_2^2 + q_2^2}{4} + \frac{1}{2} \right) \\
&\quad + X_{23} \left(P - \frac{p_1^2 + q_1^2}{2} - \frac{p_2^2 + q_2^2}{4} + \frac{1}{2} \right) \left(\frac{p_2^2 + q_2^2}{2} + \frac{1}{2} \right) \\
&\quad + K_{1,0,-1} q_1 \sqrt{2 \left(P - \frac{p_1^2 + q_1^2}{2} - \frac{p_2^2 + q_2^2}{4} \right)} \\
&\quad \times \left[1 + \lambda_1 \left(\frac{p_1^2 + q_1^2}{2} \right) + \lambda_2 \left(\frac{p_2^2 + q_2^2}{2} \right) \right] \\
&\quad + K_{1,-2,0} \frac{1}{\sqrt{2}} (q_1 q_2^2 - q_1 p_2^2 + 2 p_1 p_2 q_2) \\
&\quad + K_{0,2,-1} \sqrt{P - \frac{p_1^2 + q_1^2}{2} - \frac{p_2^2 + q_2^2}{4}} (q_2^2 - p_2^2) \\
&\quad + K_{2,0,-2} (q_1^2 - p_1^2) \left(P - \frac{p_1^2 + q_1^2}{2} - \frac{p_2^2 + q_2^2}{4} \right)
\end{aligned}
$$

For each level, 200 maximal Lyapunov exponents with randomly chosen points in the phase space were calculated and averaged. This is because that different initial points in the phase space may lead to different (maximal) Lyapunov exponents. Averaging of the exponents is to avoid bias. We call it the averaged Lyapunov exponent. (See Appendix of Chapter 3, Section 5.8 and 7.3)

8.2   State dynamics of DCO

For a better exploration of the DCO vibrational dynamics, we need to calculate the dynamical potentials[6], which are the effective potentials D-C stretching, the bending and C-O stretching experience, dynamically. This is done by varying (p_1, p_3) to obtain extreme energies E_+ (maximal) and E_- (minimal) for each (q_1, q_3) under the constraint $p_1^2 + q_1^2 + p_3^2 + q_3^2 \leq 2P$ (This will ensure positive n_2). $E_+(q_1, q_3)$ and $E_-(q_1, q_3)$ thus define the dynamical potential in which the levels sharing the common P reside. Alternatively, we can have $E_+(q_1, q_2)$ and $E_-(q_1, q_2)$.

For each level, we will simulate the stability of D-C bond by choosing randomly 200 initial points in the phase space and following their trajectories for a duration of 10 ps. D-C bond is considered as stable (*non*dissociative), if $n_{D-C} < 4$ during this time interval. The dissociation is confirmed once a trajectory has crossed the dividing surface in the direction of products and never returns to the reactant region of phase space. The percentages of nondissociation cases for the 28 levels under $P = 6$ are shown in Fig. 8.1(b). (Only the case of $P = 6$ is shown for demonstration. Those under various P show similar behaviors.) Shown in Fig. 8.1(a) are the levels that were concluded from the experiment[1,2,7], in which solid and dotted lines are the resonance (bound) and scattering (dissociative) states, respectively. The broken lines show those that are stable by our calculation (the percentage of nondissociation cases is larger than 50%) but not reported in the experiment, due to their too large resonance widths. The criterion that $n_{D-C} > 4$ is set for dissociation is based on the consideration that it coincides with the threshold region of the dissociation of D atom[3]. With this action, D-C bond has a displacement over 72% of its equilibrium bond length which is large enough for the

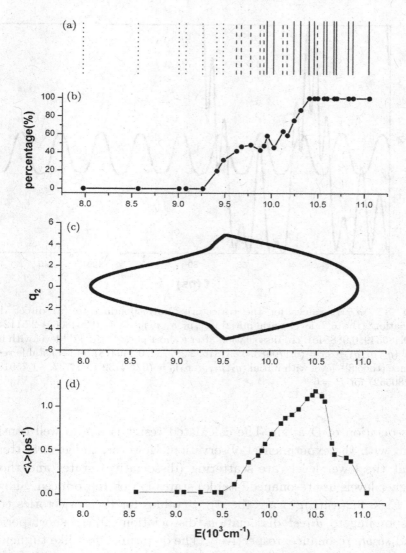

Fig. 8.1 (a) The 28 levels with $P = 6$. Solid (——) and dotted (......) lines show the resonance and scattering states, respectively. The broken (-----) lines show those that are resonant, but not reported in the experiment. (b) The percentages of nondissociative cases for these levels. (c) The dynamical potential corresponding to the bending coordinate. (d) The averaged Lyapunov exponents, $\langle \lambda \rangle$, for the levels. The level energies are with respect to the zero point energy which is $2380.075 \, \text{cm}^{-1}$ above the potential well bottom.

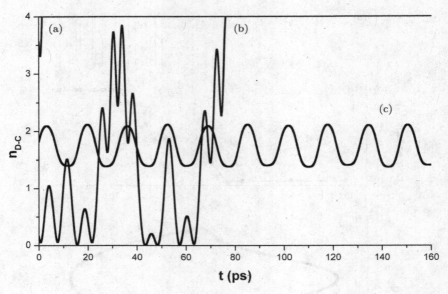

Fig. 8.2 n_{D-C} values for the trajectories corresponding to (a) direct dissociation (the 6th level with initial $(q_1, p_1, q_3, p_3)_0 = (-0.234806, -2.511240, -0.485349, 0.988546)$), (b) dissociation after a short period (the 17th level with initial $(q_1, p_1, q_3, p_3)_0 = (0.170583, 0.537316, 2.287734, 0.446478)$) and (c) stable resonance (the 28th level with initial $(q_1, p_1, q_3, p_3)_0 = (0.084182 \; 1.904737, -1.759188, 2.280552)$ for $P = 6$.

dissociation of D atom. The calculated result is not so well consistent with the experimental observation. However, the general trend that the lower levels are scattering (dissociative) states and those higher levels are resonance (stable) states is well reproduced. Shown in Figs. 8.2(a),(b),(c) are the n_{D-C} values for the trajectories corresponding to *direct* dissociation, dissociation after a short period and *stable* resonance, respectively. The quantum effect like tunneling can be crucial in the dissociation of D-C bond. This is lack in our classical treatment. This judges that our classical approach cannot well reproduce the experimental observation. However, this is not the whole story. Shown in Fig. 8.1(c) is the dynamical potential corresponding to the bending coordinate. It is noted that the energy (around 9 500 cm^{-1}) at which the bending coordinate reaches its full range corresponds well to the level that ceases dissociating. This observation is also true for the cases with other P. For instance, for small $P = 3$, there are no scattering states. Indeed, for this case, the dynamical potential of the bending coordinate does reach its full range in the very lowest energy level. For a further analysis, the averaged Lyapunov exponents for the levels are shown in Fig. 8.1(d). It is seen that the level of which the exponent starts uprising (nonzero) corresponds well to where the bending coordinate reaches its full range. This demonstrates that the formation of the resonance state and the appearing of nonzero (averaged) Lyapunov exponent are well correlated to the mediation by the bending motion. This is logical in the sense that the bending motion plays a pumping role for the energy exchange between the two stretching motions in this three body dynamics. The energy flow mediated by the bending motion between the D-C and C-O stretchings helps the relaxation of energy aggregated in the D-C bond and thus reduces its chance of dissociation. This mediation by the bending motion also helps the intramolecular vibrational relaxation and thus enhances the degree of dynamical chaos. For higher levels, we can image that, due to the stronger couplings among the three modes, the dynamics turns out to be more stable and regular. This leads to the decrease of the degree of chaos. The calculated averaged Lyapunov exponent does show this trend as demonstrated in Fig. 8.1(d). This is also confirmed in the SEP spectra[2] that the very highest levels under various P, usually possess very narrow spectral bandwidths (around 1 cm^{-1}, as contrasted with the nearby spectral bandwidths which are of 10 cm^{-1}) as shown in Table 8.1. These levels also possess very small Lyapunov exponents. With these assessments, we predict that there are relatively stable levels with narrow resonance widths at 17 942 cm^{-1} ($P = 10$), 19 611 cm^{-1} ($P = 11$) and 21 259 cm^{-1} ($P = 12$), though they have not been reported in the experiment[2].

It is impressive that the main character of this DCO vibrational dynamics can be described adequately by this classical approach.

Table 8.1   The very highest levels under various P, which possess very narrow spectral bandwidths (around $1\,\mathrm{cm}^{-1}$) and small averaged Lyapunov exponents.

Energy/ cm^{-1}	P	Level number in the energy series	Resonance width/cm^{-1}	Lyapunov exponent/ps^{-1}
9 099	5	20	0.2	0.258
9 251	5	21	0.29	0.022
10 710	6	25	0.96	0.458
10 817	6	26	0.5	0.420
10 885	6	27	0.64	0.380
11 034	6	28	0.53	0.022
12 795	7	36	0.61	0.110
14 535	8	45	0.53	0.083
16 263	9	55	0.9	0.023

8.3   Contrast of the dynamical potentials of D-C and C-O stretchings

It is interesting to note the contrast of the dynamical potentials of D-C and C-O stretchings. We show them in Figs. 8.3(a),(b), respectively. The transformation from q_i to cartesian Δr_i is via $\Delta r_i = a_i^{-1}\ln\{[1 - (1 - \lambda_i^2)^{1/2}\mathrm{sgn}]/\lambda_i^2\}$, with $q_i < 0$, sgn $= 1$; $q_i > 0$, sgn $= -1$. $\lambda_i = 1 + X_{ii}(q_i^2 + 1)/\omega_i$ and a_i^{-1} is the characteristic length of a Morse potential (See Section 1.2). The dynamical potential of D-C stretching is an anti-Morse (upside-down) potential (See Section 2.3) while that of C-O is, more or less, like an harmonic one. We, therefore, simulate the Lyapunov exponent calculation by a model of an anti-Morse oscillator coupled to an harmonic one. The Hamiltonian and Hamilton's equations of motion are:

$$v^2/2 + D[1 - \exp(-ar)]^2 + p_x^2/2 + \omega^2 x^2/2 + kp_x v$$

and

$$dr/dt = v + kp_x$$

$$dv/dt = -2aD[\exp(-ar) - \exp(-2ar)]$$

$$dx/dt = p_x + kv$$

$$dp_x/dt = -\omega^2 x.$$

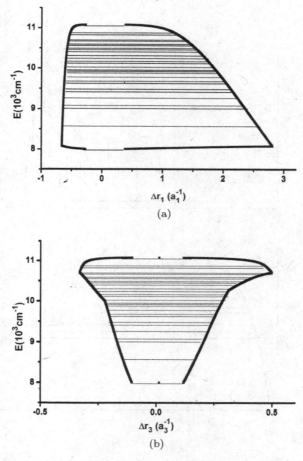

Fig. 8.3   The dynamical potentials of (a) D-C and (b) C-O stretchings for $P = 6$. The horizontal lines are the associated levels.

In the above expressions, v is the momentum of the anti-Morse oscillator. p_x is that of the harmonic oscillator. $D < 0$ and a are the parameters for the anti-Morse oscillator. ω is the angular frequency of the harmonic oscillator. k is the momentum coupling strength. For numerical simulation, we set $D = -85$, $a = 0.1$, $\omega = 10.0$, $k = 10.0$. Figures 8.4(a),(b) show the calculated Lyapunov exponents with initial $(r, v, x, p_x)_0 = (-7.0, v, 0.1, 0.02)$ and the corresponding

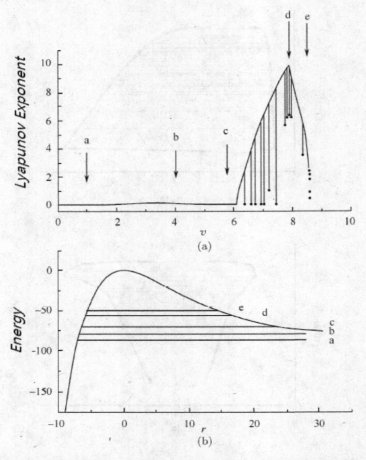

Fig. 8.4 The Lyapunov exponent calculation by a model of an anti-Morse oscillator coupled to an harmonic one. (a) The calculated Lyapunov exponents and (b) the corresponding energy levels in the anti-Morse potential.

levels in the anti-Morse potential. The reproducibility of the apparent profile of the averaged Lyapunov exponents of the DCO levels by this model is obvious except that there are *dips* in the model simulation.

8.4 The HCO case

The situation of HCO is similar to DCO. The coefficients of its algebraic Hamiltonian are obtained by fitting to the experimental observation[8,9]. The standard deviation is $7.4\,\mathrm{cm}^{-1}$. The determined coefficients are: $\omega_1 = 2858.5\,\mathrm{cm}^{-1}$, $\omega_2 = 1118.8\,\mathrm{cm}^{-1}$, $\omega_3 = 1892.8\,\mathrm{cm}^{-1}$, $X_{11} = -194.1\,\mathrm{cm}^{-1}$, $X_{12} = -46.3\,\mathrm{cm}^{-1}$, $X_{13} = 4.2\,\mathrm{cm}^{-1}$, $X_{22} = -8.2\,\mathrm{cm}^{-1}$, $X_{23} = -4.9\,\mathrm{cm}^{-1}$, $X_{33} = -11.9\,\mathrm{cm}^{-1}$, $K_{101} = 38.7\,\mathrm{cm}^{-1}$, $K_{120} = 10.7\,\mathrm{cm}^{-1}$, $K_{021} = -3.3\,\mathrm{cm}^{-1}$, $K_{202} = 9.4\,\mathrm{cm}^{-1}$, $\lambda_1 = 0.2$, $\lambda_2 = -0.3$. Compared with the DCO system, these parameters are reasonable. For instance, ω_1 for HCO is 1.4 fold that of DCO, due to the isotope effect. ω_2 for HCO is also larger than that for DCO. ω_3's for the two systems are quite similar. X_{11} and X_{22} for HCO are larger. Furthermore, the Fermi resonance strength for HCO is smaller since its $\omega_1 - 2\omega_2$ is larger than that for the DCO system.

Similar to those for DCO, the percentages of nondissociative cases, the dynamical potentials along the bending coordinate and the averaged Lyapunov exponents for $P = 6, 8$ and 10 are shown in Figs. 8.5(a),(b),(c). In the figures, also shown are the spectral widths for some levels which are available from Ref. [8]. It is noted that for the levels sharing a common polyad number, the energy at which the bending coordinate (q_2) of the dynamical potential reaches its full range corresponds well to the level which possesses a significantly dropping spectral width. (The narrower the spectral width is, the less the probability that H-C dissociates.) Significantly, this is also the region where the averaged Lyapunov exponent enhances greatly. Specifically, for low levels, excitation tends to the dissociative/scattering state with wider spectral width. For high levels, bending motion between the H-C and C-O stretchings helps the relaxation of energy aggregated in the H-C bond and thus reduces the chance of its dissociation. This leads to a narrower spectral width. IVR is mediated mainly by the bending motion. This leads to the enhancement of the degree of chaos and the increase of the Lyapunov exponent.

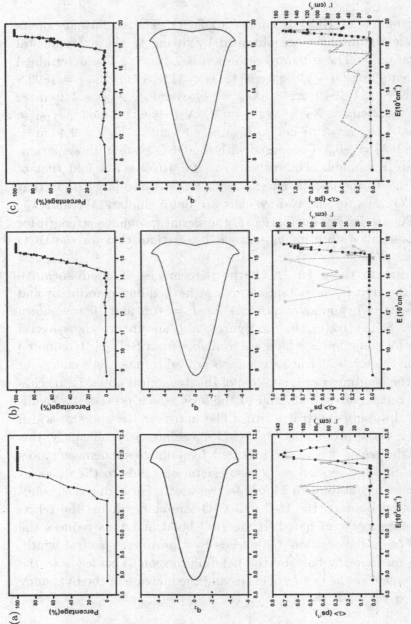

Fig. 8.5 For the levels with polyad number (a) $P = 6$, (b) $P = 8$, (c) $P = 10$, the percentages of nondissociative cases, the dynamical potentials along the bending coordinate (q_2) and the averaged Lyapunov exponents, $\langle\lambda\rangle$ denoted by (■). Also shown are the band widths Γ (denoted by ○) available from Ref. [8].

8.5 Comparison of the dynamical potentials

Shown in Fig. 8.6 are the dynamical potentials along q_1 (the H–C stretching), q_2 (the H-C-O bending) and q_3 (the C-O stretching) coordinates for HCO with $P = 6$. The horizontal segments are the corresponding levels. It is shown that all the levels are enclosed in the dynamical potentials. Furthermore, it is interesting to note the anti-Morse type for the dynamical potential along the H-C stretching coordinate as contrasted to those along the bending and C-O stretching coordinates which are of Morse type in the lower energy part. We note that for a Morse potential, the lowest point is stable and the highest point is unstable corresponding to dissociation while for an anti-Morse potential, the situation is just *up-side-down* that the highest point is a stable point and the lowest point is unstable corresponding to dissociation. Therefore, for this type potential, lower levels may be prone to dissociation while higher levels are the more stable resonance states. Apparently, this is related to the dissociative nature of the H-C bond. In summary, we conclude that dynamical potentials do show the dynamical nature of this three body problem.

8.6 A comment: The IVR role of bending motion

The dynamics of scattering and resonance states in both DCO and HCO is expected to be quantum mechanical. However, this does not rule out the semi-classical approach from certain aspects. We demonstrated that the coset space dynamics, Lyapunov analysis and dynamical potential are useful for the understanding of state dynamics. It was also demonstrated that the level dynamics as classified by the polyad number shows consistent behaviors. This shows, dynamically, that there is regularity behind these seemingly complicated levels, if appropriate method and viewpoint are taken upon.

Rather important is that through these semi-classical analyses, the IVR role by the bending motion is clearly and consistently confirmed.

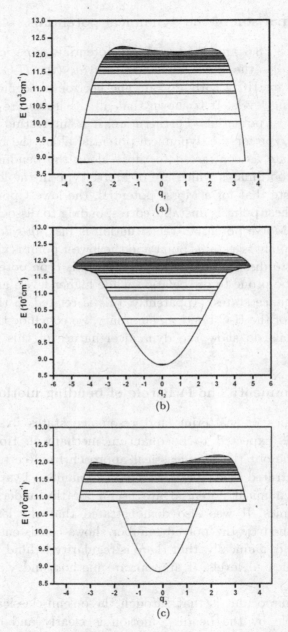

Fig. 8.6   The dynamical potentials along q_1 (the H-C stretching), q_2 (the H-C-O bending) and q_3 (the C-O stretching) coordinates for HCO with P = 6. The horizontal segments are the corresponding levels.

8.7 Dynamical barrier due to extremely irrational couplings: The role of bending motion

In nonstatistical IVR, there are bottleneck barriers that block the energy transfer among the modes. Martens *et al.*[10] have proposed that this is related to the extremely irrational ratios among the classical vibrational frequencies in their concept of frequency ratio space and resonance lines. The so-called extremely irrational numbers are $(\sqrt{5} - 1)/2 \pm integers$. Hashimoto and Someda[11] have observed that IVR duration between the two O-H stretchings (over 1 ps) is much longer than that between the stretching and bending (in 100 fs) in H_2O. This hints that direct vibrational energy flow between the two O-H stretchings is hindered. In this section, it will be demonstrated that there is a region in the dynamical phase space in which resonance and barrier spaces overlap, though they are distinctly two spaces. Moreover, the barrier region is shown to be composed of periodic and/or quasiperiodic motions which, therefore, block the energy transfer. The Chirikov conjecture (See Section 3.4) that overlapping of resonances leads to chaos will also be demonstrated by a simulation.

The three-mode case of H_2O will be employed to analyze the detailed intramolecular dynamical properties due to extremely irrational couplings. The physical significance is that barrier is more on the energy flow between the two stretchings, though there is transfer between one stretching and bending. In contrast, resonance is, more or less, mediated by the bending mode, i.e., through Fermi resonance. This will certainly lead to dynamical chaos, or statistical IVR.

The H_2O Hamiltonian is shown in Section 5.4 and the coset representation is in Section 2.2. The coupling condition is defined by $K \cdot \omega_0 = 0$ with $\omega_0 = (\omega_{0s}, \omega_{0t}, \omega_{0b})$ and $\omega_{0\alpha} = \partial H_0/\partial n_\alpha$. For the 1:1 and Fermi couplings, $K1 = (1, -1, 0)$, and $K2 = (1, 0, -2)$, $K3 = (0, 1, -2)$, respectively. For the extremely irrational couplings, we will have $K4 = (1, \gamma, -2(\gamma + 1))$ and $K5 = (1, \gamma + 1, -2(\gamma + 2))$ with $\gamma = (\sqrt{5} - 1)/2$ for our analysis. This will be explained below.

Numerical calculation is with the precision up to the 15th digit after the decimal point. In numerical calculation, we adopt the condition that $|K \cdot \omega_0| \leq 0.1$ and $P = 15$. For this action, the energy is from 38 000 cm^{-1} to 52 000 cm^{-1}, which is high enough. Of course, we may adopt higher precision, but the time consumption is beyond our tolerance. In fact, we found that the analyses and conclusions based on either $|K \cdot \omega_0| \leq 0.1$ or $|K \cdot \omega_0| \leq 1.0$ are quite the same.

The three resonance lines in the action space due to $K1$, $K2$ and $K3$ intersect at a point. We can be assured that there are two coupling lines relating to the extremely irrational numbers: $K4 = (1, \gamma, -2(\gamma + 1))$ and $K5 = (1, \gamma + 1, -2(\gamma + 2))$ that coincide with the intersection point. For short, they are called the barrier lines. This implies that there is a dynamical region in which both the chaotic and barrier motions coexist. These lines and their intersection point in the action space are shown in Fig. 8.7(a).

For dynamical study, we randomly choose points from the resonance and barrier lines and follow their trajectories, evolutions of $\Delta \omega_0 = K_i \bullet \omega_0$ and actions as a function of time. For demonstration, only the situations resulting from two initial points in the resonance $(K1)$ and barrier $(K4)$ lines, respectively, are shown here. The demonstrations show the typical behaviors of the points in the resonance and barrier lines. The trajectories are shown in Figs. 8.7(a),(b), respectively. It is evident that the trajectories are strongly hindered by the barrier lines. The corresponding evolutions of $\Delta \omega_0$ are shown in Figs. 8.7(c),(d). The initial $\Delta \omega_0$ is zero in both cases. However, for the $K1$ case (Fig. 8.7(c)), $\Delta \omega_0$ can be positive and negative. This implies that the corresponding trajectory wanders around the resonance line, though limited by the barrier lines, as shown in Fig. 8.7(a). For the $K4$ case (Fig. 8.7(d)), $\Delta \omega_0$ can only be positive, showing that the corresponding trajectory is hindered as shown in Fig. 8.7(b). Shown in Figs. 8.7(e),(f) are the temporal actions for these two cases. For the $K1$ case (Fig. 8.7(e)), action/energy can flow freely among the two stretching and bending modes. However, for the $K4$ case (Fig. 8.7(f)), action/energy can only transfer between one stretching and bending modes. The hindrance to IVR is very obvious.

The intersecting regions by the resonance and extremely irrational (barrier) couplings, respectively, deserve attention. That by

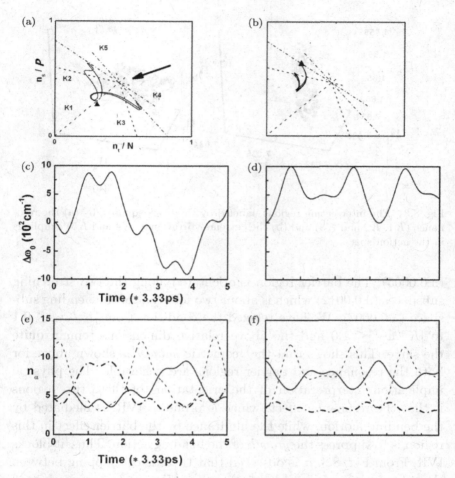

Fig. 8.7 (a) The intersection, denoted by the arrow, of the resonance ($K1$, $K2$ and $K3$) and barrier ($K4$ and $K5$) lines due to extremely irrational numbers in the action space. Also shown is the trajectory (the starting point is denoted by ▲) originating from the resonance ($K1$) line. (b) The trajectory originating from the barrier ($K4$) lines. The trajectory duration in (a) and (b) is 16.65 ps. (c),(e) and (d),(f) are their corresponding evolutions of $\Delta\omega_0 = K_i \bullet \omega_0$ and actions. In (e) and (f), the solid, dashed and dash-dotted lines are n_s, n_b and n_t, respectively.

$K1$, $K2$ and $K3$ in the action space is shown in Fig. 8.8(a) and that by $K4$ and $K5$ is shown in Fig. 8.8(b). It is interesting to note that the region by the resonances is elongated along the bending subspace (±0.0014) which is about three fold that in the stretching subspace

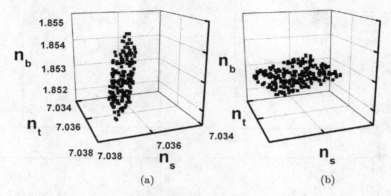

Fig. 8.8   The intersecting regions, labeled by the black squares, by (a) the resonance ($K1$, $K2$ and $K3$) and (b) the extremely irrational ($K4$ and $K5$) couplings in the action space.

(± 0.0005). The barrier region is orientated along the two stretching subspaces (± 0.0012) which is about two fold that of the bending subspace (± 0.0007). We have relaxed the condition from $|K \cdot \omega_0| \leq 0.1$ to $|K \cdot \omega_0| \leq 1.0$ and the above relative dimensions remain quite the same. This shows that the geometric *shapes* as shown above for both the resonance and barrier regions are believable. The physical implication/interpretation of the orientations of these two regions is that the resonance effect, which is prone to IVR, is mediated by the bending motion, while the hindrance by the barrier effect in this region is to suppress the *growth* of the bending action. Thus, it blocks IVR. From Fig. 8.8, it is observed that there is overlapping between these two (resonance and barrier) regions. However, they are indeed two distinct spaces. Shown in Fig. 8.9(a),(b) are the two trajectories originating from these two spaces, respectively. Again, it is observed that the trajectory from the resonance space can wander more freely than that from the barrier space, which is rather restricted by the barrier lines as evidenced in the action space.

The main picture is that the irrational couplings impose hindrance to IVR and act as barriers. They block direct action/energy transfer between the two O-H stretchings, though allow the transfer between one stretching and bending motions. Resonance couplings which lead to chaos of IVR are mainly mediated by the bending

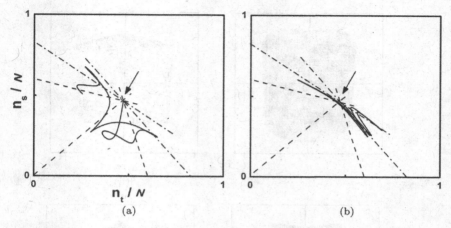

Fig. 8.9   The two trajectories (the starting points are denoted by the arrows), originating from the intersecting regions, by (a) the resonance ($K1$, $K2$ and $K3$) and (b) the extremely irrational ($K4$ and $K5$) couplings as shown in Fig. 8.8.

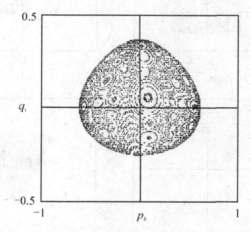

Fig. 8.10   The barrier region of the 30th level of the system of H_2O, with $P = 15$.

mode. There is also a dynamical region in which resonance and extremely irrational couplings coexist.

Figure 8.10 shows the barrier region of the 30th level of the system of H_2O, with $P = 15$. It is seen that the barrier is composed of periodic and quasiperiodic motions. The periodic and quasiperiodic

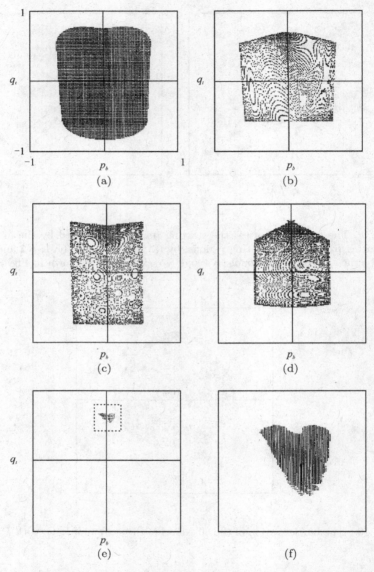

Fig. 8.11 The phase space structure and resonance regions in (q_t, p_b) for the 116th state of H_2O with $P = 15$. (a) is the solution space by $E = H(q_t, p_t, q_b p_b)$. (b) is due to 1:1 and 2:2 resonances. (c), (d) are due to Fermi resonance. (e) is the overlap of (b), (c),(d). (f) is the enlargement of (e).

motions cannot deliver energy out via the chaotic motion. Hence, they stop the energy transfer.

In addition, to demonstrate the conjecture by Chirikov (See Section 3.4), we show in Fig. 8.11(a) the solution space by $E = H$ (q_t, p_t, q_b, p_b). Figure 8.11(b) is due to 1:1 and 2:2 resonances. Figures 8.11(c),(d) are due to Fermi resonance. Figure 8.11(e) is the overlap of Figs. 8.11(b),(c),(d). Figure 8.11(f) is the enlargement of Fig. 8.11(e). From Figs. 8.11(b)–(d), we see that resonance regions have a complicated structure, composed of many periodic, quasiperiodic motions and a small chaotic region. Also confirmed is that the overlapping region is full of chaos as conjectured by Chirikov.

References

1. Stock C, Li X, Keller H M, Schinke R, Temps F. *J. Chem. Phys.*, 1997, 106: 5333.
2. Keller H M, Stumpf M, Schroder T, Stock C, Temps F, Schinke R, Werner H J, Bauer C, Rosmus P. *J. Chem. Phys.*, 1997, 106: 5359.
3. Troellsch A, Temps F. *Z. Phys. Chem.*, 2001, 215: 207.
4. Stamatiadis S, Faratos S C, Keller H M, Schinke R. *Chem. Phys. Lett.*, 2001, 344: 565.
5. Jung C, Taylor H, Atilgan E. *J. Phys. Chem. A*, 2002, 106: 3092.
6. Kellman M E. *J. Chem. Phys.*, 1985, 83: 3843.
7. Keller H M, Schinke R. *J. Chem. Phys.*, 1999, 110: 9887.
8. Keller H M, Floethmann H, Dobbyn A J, Schinke R, Werner H J, Bauer C, Rosmus P. *J. Chem. Phys.*, 1996, 105: 4983.
9. Tobiason J D, Dunlop J R, Rohlfing E A. *J. Chem. Phys.*, 1995, 103: 1448.
10. Martens C C, Davis M J, Ezra G S. *Chem. Phys. Lett.*, 1987, 142: 519.
11. Hashimoto N, Someda K. *Chem. Phys. Lett.*, 2000, 323: 79.

Chapter 9

Dynamical Potential Analysis for HCP, DCP, N₂O, HOCl and HOBr

9.1 Introduction

In our previous work (Chapter 8) on DCO and HCO, the dynamical potential analysis was successfully employed to study the dissociation of D–C and H–C bonds. The bending induced isomerization of HCP (phosphaethyne) to HPC is a challenging topic. The reason is that: (1) this is an elementary chemical mechanism, (2) enough experimental data have been acquired. (3) theoretical works by *ab initio* potential energy surface calculation, by both semi-classical and wave function algorithms and by nonlinear mechanical concepts including bifurcation, have acquired significant results.

Both HCP and HPC are linear. HCP is stable. HPC is unstable. It is known that H–C stretching is rather decoupled from C–P stretching and H–C–P bending motions. The latter two possess Fermi resonance. At high excitation, strong inter-mode coupling will lead to strong IVR (Intramolecular Vibrational Relaxation) that the concept of forming HPC by exciting H–C–P bending motion high enough may not be realistic. However, the situation is not so straight forward. There is possibility that nonlinear effect at high excitation will result in the formation of localized bending mode. Thus the isomerization via H–C–P bending motion can come true.

The works in the literature are mostly based on rather complicated computation. Here, we will demonstrate that by the coset represented Hamiltonian, its dynamical potentials can be elucidated easily, from which the fixed points can be identified immediately. We will recognize that fixed points *define* the dynamical properties in a *global* sense, from both the classical and quantal viewpoints. Thereby, the localized H–C–P bending mode can be easily identified.

Other works on DCP, N_2O, HOCl and HOBr will also be shown. The advantage of this approach is indeed convincing.

9.2   The coset represented Hamiltonian of HCP

The algebraic Hamiltonian of HCP including its coefficients have been determined in the literature.[1] For notation, the subscripts, 1, 2 and 3 are used for H–C stretching, H–C–P bending and C–P stretching. The Hamiltonian is:

$$
H(n_1, n_2, n_3) = \omega_1 \left(n_1 + \frac{1}{2} \right) + \omega_2 (n_2 + 1) + \omega_3 \left(n_3 + \frac{1}{2} \right)
$$

$$
+ X_{11} \left(n_1 + \frac{1}{2} \right)^2 + X_{12} \left(n_1 + \frac{1}{2} \right) (n_2 + 1)
$$

$$
+ X_{13} \left(n_1 + \frac{1}{2} \right) \left(n_3 + \frac{1}{2} \right) + X_{22} (n_2 + 1)^2
$$

$$
+ X_{23} (n_2 + 1) \left(n_3 + \frac{1}{2} \right) + X_{33} \left(n_3 + \frac{1}{2} \right)^2
$$

$$
+ y_{222} (n_2 + 1)^3 + z_{2222} (n_2 + 1)^4
$$

$$
- \left[k + k_1 \left(n_1 + \frac{1}{2} \right) + k_2 (n_2 + 2) + k_3 n_3 \right]
$$

$$
\times (a_2^2 a_3^+ + a_2^{+2} a_3)
$$

where n_i are the actions and the coefficients are: $\omega_1 = 3343.1225$ cm^{-1}, $\omega_2 = 697.7797$ cm^{-1}, $\omega_3 = 1301.0838$ cm^{-1}, $X_{11} = -55.0161$ cm^{-1}, $X_{12} = -16.8174$ cm^{-1}, $X_{13} = -4.3375$ cm^{-1},

$X_{22} = -5.3477$ cm^{-1}, $X_{23} = -4.6460$ cm^{-1}, $X_{33} = -5.8619$ cm^{-1}, $y_{222} = 0.23345$ cm^{-1}, $z_{2222} = -0.00562$ cm^{-1}, $k = 3.6115$ cm^{-1}, $k_1 = 0.8056$ cm^{-1}, $k_2 = 0.06727$ cm^{-1}, $k_3 = -0.22067$ cm^{-1}. (The fit is by approaching the eigenenergies of this Hamiltonian with quantized actions to those levels obtained from the exact quantum calculation.) The characteristics of this system is that nonlinearity and Fermi resonance are eminent. The coupling of H–C stretching and H–C–P bending is very weak. The conserved action is $P = n_2 + 2n_3$, besides n_1. P will be limited to 32 due to that these coefficients were determined by fitting to the level energies with $P < 32$.

The coset represented Hamiltonian is accomplished through the substitution, $n_2 = (q_2^2 + p_2^2)/2$, with constants n_1 (which will be set to 0 in our case study) and P. Thus, $n_3 = [P - (q_2^2 + p_2^2)/2]/2$ and

$$
\begin{aligned}
H(q_2, p_2) &= \frac{1}{2}\omega_1 + \omega_2\left(\frac{p_2^2 + q_2^2}{2} + 1\right) + \omega_3\left(\frac{P}{2} - \frac{p_2^2 + q_2^2}{4} + \frac{1}{2}\right) \\
&\quad + \frac{1}{4}X_{11} + \frac{1}{2}X_{12}\left(\frac{p_2^2 + q_2^2}{2} + 1\right) \\
&\quad + \frac{1}{2}X_{13}\left(\frac{P}{2} - \frac{p_2^2 + q_2^2}{4} + \frac{1}{2}\right) \\
&\quad + X_{22}\left(\frac{p_2^2 + q_2^2}{2} + 1\right)^2 + X_{23}\left(\frac{p_2^2 + q_2^2}{2} + 1\right) \\
&\quad \times \left(\frac{P}{2} - \frac{p_2^2 + q_2^2}{4} + \frac{1}{2}\right) + X_{33}\left(\frac{P}{2} - \frac{p_2^2 + q_2^2}{4} + \frac{1}{2}\right)^2 \\
&\quad + y_{222}\left(\frac{p_2^2 + q_2^2}{2} + 1\right)^3 + z_{2222}\left(\frac{p_2^2 + q_2^2}{2} + 1\right)^4 \\
&\quad - \left[k + \frac{1}{2}k_1 + k_2\left(\frac{p_2^2 + q_2^2}{2} + 2\right)\right. \\
&\quad \left. + k_3\left(\frac{P}{2} - \frac{p_2^2 + q_2^2}{4}\right)\right]\sqrt{\frac{P}{2} - \frac{p_2^2 + q_2^2}{4}}(q_2^2 - p_2^2).
\end{aligned}
$$

Similarly, we can have $H(q_3, p_3)$. For short, it is shown in the Appendix.

9.3   Dynamical potentials and state properties inferred by action population

The dynamical potential of $H(q_i, p_i)$ with P given is the *effective environment* in which the q_i coordinate experiences. This is achieved by calculating the maximal and minimal energies by varying p_i for each q_i under the condition that actions are nonnegative. The dynamical potential composed of these maximal and minimal energies as a function of q_i is a closed curve in which all the quantized levels possessing the same P will reside. The points on the dynamical potential curve correspond to $\partial H / \partial p_i = 0$. The dynamical potentials for HCP with $P = 22$ are shown in Figs. 9.1(a) and (b) in the q_2 and q_3 coordinates, in which the horizontal lines show the energy levels sharing the designated P. The case of $P = 22$ will be singled out for discussion since it is quite representative and possesses the most fruitful characteristics. The level energy is with respect to the ground level which is 2987.72 cm^{-1} above the bottom of the potential well. Fixed points satisfy the conditions that both its $\partial H / \partial q_i$ and $\partial H / \partial p_i$ $(i = 2$ or 3$)$ are zero. As time evolves, these fixed points remain unmoved. Their corresponding actions $(n_2/2$ or $n_3)$ are also constants. We have to stress that the points in Fig. 9.1 designated by $[B]$, $[r]$, $[SN]$ are stable fixed points and $[\overline{SN}]$ is an unstable fixed point. We note that $[B]$ is stable since for the upper realm of the dynamical potential, we have to interpret the potential curve in an inverse (upside down) way. $[r]$ in Fig. 9.1(b) is an interval with q_3 ranging horizontally roughly from -6 to 6. The unstable $[\overline{SN}]$ in Fig. 9.1(b) is an inflection point where the second derivative with respect to q_3 is zero.

For each level, we can calculate actions, $n_2/2$ and n_3, from $q_2(q_3)$ and $p_2(p_3)$ by equating its energy to $H(q_2, p_2)$ $(H(q_3, p_3))$. Then, we can have the relation of $n_2/2$ (n_3) against $q_2(q_3)$. This is called the action population. Action population is similar to the probability distribution derived from wave function. These quantities as a function of q_2 for all the 12 levels corresponding to $P = 22$ are shown in Fig. 9.2.

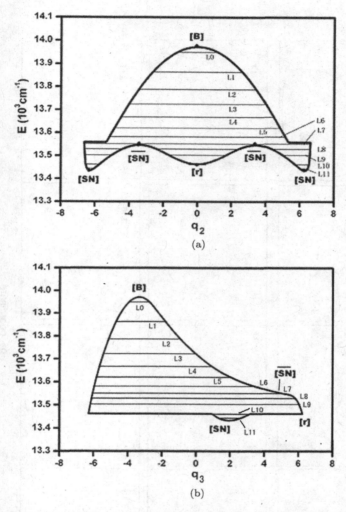

Fig. 9.1   The dynamical potentials in q_2 (a) and q_3 (b) coordinates for HCP with $P = 22$. The horizontal lines show the quantal levels, L0 to L11. $[B]$, $[r]$, $[SN]$ and $[\overline{SN}]$ are the stable and unstable fixed points, respectively.

For these levels, we have the following interpretations:

L0 ($13\,950$ cm^{-1}): It is close to $[B]$. Since its $n_2/2$ is always larger than n_3, the main motion is HCP bending. However, it is different from L11 (see below). L0 has less component in the HCP bending motion than L11.

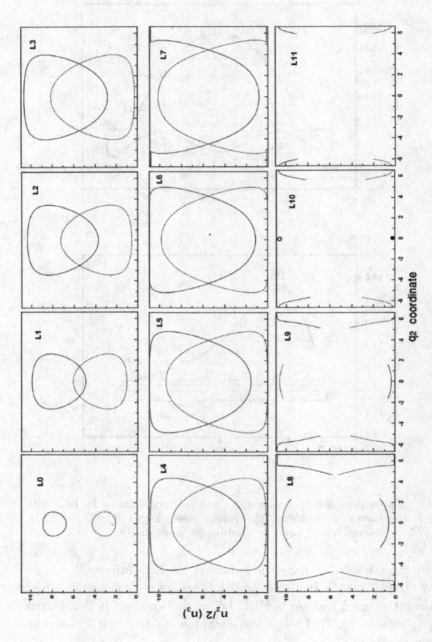

Fig. 9.2   The actions, $n_2/2$ and n_3 as the functions of q_2 coordinate for the levels of HCP with $P = 22$. Solid line and circle are for $n_2/2$. Broken line and open circle are for n_3.

L1–L7 (The level energies are 13 863 cm^{-1}, 13 788 cm^{-1}, 13 723 cm^{-1}, 13 668 cm^{-1}, 13 621 cm^{-1}, 13 582 cm^{-1}, 13 551 cm^{-1}): All these levels have a connected region for q_2 in which $n_2/2$ can be larger or less than n_3. (Note that in the Figure, $n_2/2 + n_3$ is a constant.) This indicates that IVR is strong between HCP bending and C–P stretching motions. Furthermore, the action population structure of L7 is significantly different from those lower levels. This indicates a dynamical bifurcation.

L8 (13 530 cm^{-1}) and L9 (13 504 cm^{-1}): In contrast to the cases of L1 to L7, there are two segregated regions (q_2), in which $n_2/2$ is larger and less than n_3, respectively. This means that action distributions in the bending and C–P stretching are uneven and hence IVR is weaker.

In our analysis, if the motion starts with smaller and larger $|q_2|$, respectively, for L8 and L9, then the motion will mainly focus on the C–P stretching for L8 and bending for L9. This is in agreement with that concluded in Ref. [1]. (In fact, L8 possesses larger region in q_2. It will have more probability in C–P stretching than L9 if quantum mechanically interpreted.) However, if the motion starts with the otherwise q_2 regions, the motion will mainly focus on the bending for L8 and C–P stretching for L9. Then this is a variation. In fact, albeit the wave functions of L8 and L9 shown in Ref. [1] imply mainly that L8 possesses more C–P stretching while L9 more bending motion, one cannot rule out their motions, respectively, along the bending and C–P stretching though with less probability. In this case, we have to appreciate the power of this classical dynamical analysis for its simplicity.

L10 (13 462 cm^{-1}) This level is on the stable fixed point at $q_2 = 0$ ([r] in Fig. 9.1) which corresponds to the pure C–P stretching motion. (See Section 9.8.) This level will remain in this motion forever if it starts there. This is what shown by the quantal calculation. Again, there is the situation that this level is full of the bending motion as it stays in the larger $|q_2|$ region.

L11 (13 445 cm^{-1}): This level mainly possesses the HCP bending motion with very tiny portion of the C–P stretching. IVR is very weak between HCP bending and C–P stretching. Hence, this mode can be

the most potential one for the isomerization from HCP to HPC, if it
possesses enough energy. The highest level L0, though also mainly of
the HCP bending, is different from L11 in that its IVR between the
HCP bending and C–P stretching motions is stronger and it contains
more C–P stretching character. The bending action (energy) of L0 is
just less than that of L11, though the energy of L0 is higher than that
of L11 (See Section 9.5). Hence, L0 is unable to convey more energy
than L11 toward the formation of HPC. Since L11 possesses very
similar amount of energy and character to the stable fixed point $[SN]$,
we call this very localized bending mode (state) which is of little IVR
and C–P stretching, the $[SN]$ mode (state). (However, L11 is different
from the fixed point [SN], dynamically. This difference cannot be
overlooked.) We will address more on this mode in Section 9.5.

Shown in Fig. 9.3 are the wave function probabilities of these
levels adopted from Ref. [1]. Our result indicates that L0 possesses

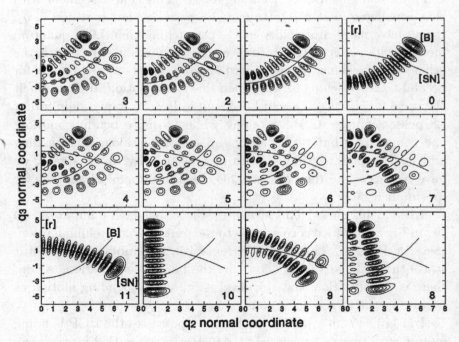

Fig. 9.3   The wave function probability, adopted from Ref. [1], of the 12 levels
with $P = 22$. q_2, q_3 are the bending and C–P stretching coordinates, respectively.

mainly the HCP bending motion and has action population character different from those of the levels from L1 to L7. This is consistent with that shown in Fig. 9.3. Our result also shows that the action populations for the levels below and above L7 are different and that there is more IVR for levels higher than L7. The probability densities shown in Fig. 9.3 also demonstrate this distinction and differentiation.

Figure 9.3 shows that L10 and L11 are mainly of C–P stretching and bending motions, respectively. This is consistent with Fig. 9.2. Figure 9.3, unlike Fig. 9.2, shows that L9 possesses more bending motion, definitely. Hence, Ref. [1] classifies L9, like L11 as localized bending mode and calls it the isomerization state, indicating that it is the state that can lead to HPC. This classification of L9 is not so evident in our analysis by Fig. 9.2. Apparently, there is definition variation between ours and Ref. [1]. As far as the definition of localization is concerned, ours is stricter.

Overall, our analysis is mostly consistent with that by the quantal wave function calculation. However, our classical treatment shows its simplicity with clear physical picture.

9.4   State classification and quantal environments

We can calculate the trajectories, which are closed contours in the (q_2, p_2) and (q_3, p_3) coset spaces, for each level of HCP by setting its eigenenergy to $H(q_2, p_2)$ or $H(q_3, p_3)$. For demonstration, only the (q_2, p_2) contours for the 12 quantal levels with $P = 22$ are shown in Fig. 9.4. Their corresponding action integrals:

$$1/2\pi \oint p_i dq_i$$

in the (q_2, p_2) and (q_3, p_3) spaces can also be calculated. The action integrals for all the levels corresponding to $P = 22$ are tabulated in Table 9.1. (We note that the action integral defined here is different from n_i shown in the previous section.)

According to the action integrals, these levels can be categorized into three classes: L0 to L7, L8 to L10 and L11. We note that the differences in action integrals between neighboring levels are quite the same, around 2. (This value could be 1 if the polyad number is re-defined as $n_2/2 + n_3$.) In the (q_2, p_2) space, the action integral from L0 to L7 is increasing. This is understandable since these levels stay in an *inverse* harmonic-like potential that for the lower levels, their action integrals are larger. For the levels from L8 to L10, their

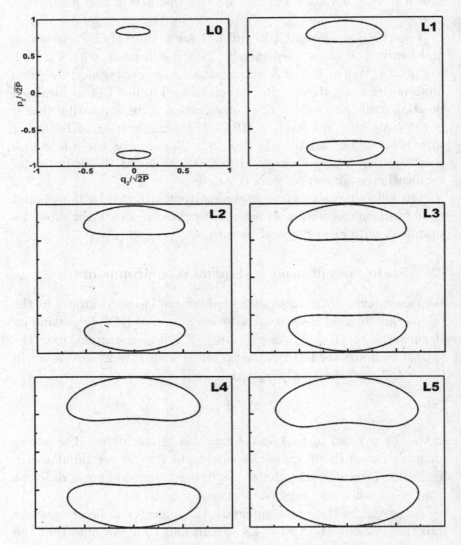

Fig. 9.4   The (q_2, p_2) contours for the 12 quantal levels with $P = 22$.

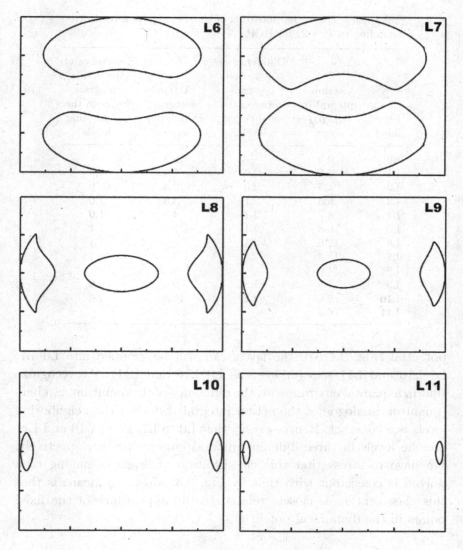

Fig. 9.4   *(Continued)*

action integrals are decreasing since they lie in a *regular* potential. L11 is, by itself, a unique state. In the (q_3, p_3) space, except L11, all the levels lie in an inverse or anti-Morse type (or anti-harmonic-like) potential. So, their action integrals increase from L0 to L10 with an almost constant increment. If one prefers, according to the dynamical

Table 9.1 The action integrals of the 12 quantal levels corresponding to $P = 22$ for HCP.

Level label	Action integral in (q_2, p_2) space	Difference of the action integrals between the neighboring levels	Action integral in (q_3, p_3) space	Difference of the action integrals between the neighboring levels
L0	0.5	—	0.2	—
L1	2.5	2.0	1.3	1.0
L2	4.5	2.0	2.3	1.0
L3	6.6	2.1	3.3	1.0
L4	8.7	2.1	4.3	1.0
L5	10.8	2.1	5.4	1.1
L6	12.9	2.1	6.5	1.1
L7	15.4	2.5	7.6	1.1
L8	4.6	—	8.7	—
L9	2.8	1.8	9.6	0.9
L10	0.6	2.2	10.8	1.2
L11	0.2	—	0.1	—

potential (Fig. 9.1(b)), the levels can still be grouped into L0–L7, L8–L10 and L11 sets (or even as L0–L10 and L11). We recognize that in a quantal environment, the difference of the quantum numbers (quantum analogue of the action integral) between the neighboring levels is a constant. Hence, we say that L0 to L7, L8 to L10 and L11 are the levels in three different quantal *environments*, respectively. We have to stress that this classification of levels belonging to a polyad is consistent with that by Fig. 9.3. Most significant is that this classification is closely related to the appearance of the fixed points in the dynamical potential.

9.5 Localized bending mode

As mentioned previously, L11 possesses the character that can lead to the formation of HPC. The action localization of this $[SN]$ mode is also obvious in other P cases. We estimate action percentages for L11 on the HCP bending and C–P stretching coordinates by integrating $n_2/2$ and n_3, respectively, through the allowed q_2 coordinate

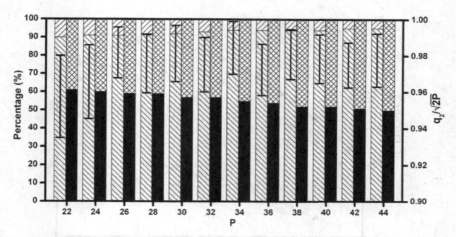

Fig. 9.5 The action percentages of $n_2/2$ (left-up shadow) and n_3 (right-up shadow) for the $[SN]$ mode as the functions of P. Bar I shows the allowed range of q_2 coordinate. This shows the localization of the $[SN]$ mode. Black and cross bars are for $n_2/2$ and n_3, respectively, of the highest levels for various P, for which IVR is evident.

as shown in Fig. 9.2. The result shows that 90% of the total action is on the HCP bending coordinate and only 10% on the C–P stretching coordinate for L11. This trend keeps on with more action on the bending coordinate as P increases. This is shown in Fig. 9.5. (P is extended to 44.)

This $[SN]$ mode is the lowest one among the levels sharing a polyad number. However, it possesses the largest portion of action on the HCP bending coordinate and is the best candidate for the formation of HPC through the bending excitation. Other levels, though with larger energies, will distribute more of its energy to the C–P stretching coordinate due to IVR. Just like the highest level, its action percentage of $n_2/2$ over that on the C–P stretching coordinate is smaller than that of the $[SN]$ mode. This is true for the other P cases as shown in Fig. 9.5. The energies of these two levels (the highest one and the $[SN]$ mode) after being multiplied by their action percentages of $n_2/2$ show that the highest level always possesses less bending energy than does the $[SN]$ mode. This is shown in Fig. 9.6. (P is extended to 44.) In Fig. 9.5, we see that the action percentage

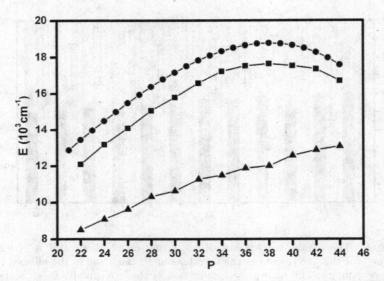

Fig. 9.6   The energy of the highest level (▲) and [SN] mode (■) after being multiplied by $n_2/2$ percentage shown in Fig. 9.5. (●) shows the [SN] mode energy.

of $n_2/2$ for the highest level decreases somewhat as P increases in contrast to that of the $[SN]$ mode which increases slightly. This means that IVR is eminent for the highest level while very little for the $[SN]$ mode.

Shown in Fig. 9.7 are the 540 levels from $P = 0$ to 44 (P is extended beyond 32), among which there are 24 $[SN]$ modes as indicated by arrows. We expect via their excitation, there is the possibility that HPC can be formed if their bending energies are enough to overcome the transition point (We will address this in Section 9.7.) From the experimental viewpoint, their identification is important. Without this identification, the experiment could be misleading.

Before conclusion, we point out one interesting phenomenon: the symmetry breaking in the lowest level for $P = 21$. This level is very similar to the lowest level for $P = 22$, but it coincides with $[r]$ and $[SN]$. Then $[r]$ and $[SN]$ share the same energy. (For other lowest levels, they are just close to, but not on both $[r]$ and $[SN]$, like here for $P = 21$.) If this level can stay on $[SN]$, then it will be mainly

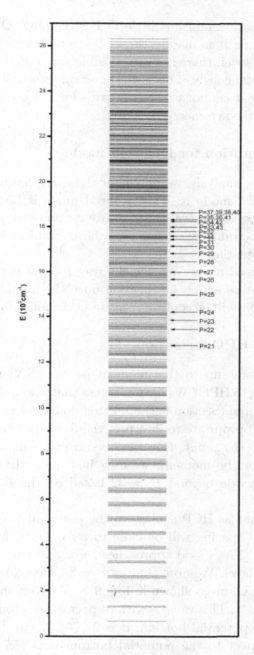

Fig. 9.7   The levels for *P* from 0 to 44. Arrows indicate the [*SN*] mode.

the bending motion and has little C–P stretching. Otherwise, if it stays on [r], then it is mainly the C–P stretching motion. In other others, for this level, there are two possible very different modes, but not both simultaneously. This is the classical interpretation. Strictly speaking, we just cannot assure that this level is on the [SN] or [r] mode. This is the symmetry breaking.

9.6 The condition for localized mode

From the above analysis, we may infer the condition for the appearance of localized mode is: The localized mode is born in the level which possesses an interval centred around nonzero q_i and lies just above/below a stable fixed point in the dynamical potential/anti-Morse type potential.

This is indeed a very convenient way for the prediction of the levels that possess action localization once the dynamical potential is available. This will be confirmed again in the following case studies.

9.7 On the HPC formation

One crucial point up to this moment is: can [SN] mode be high enough leading to HPC? We have to stress that the coefficients for the HCP Hamiltonian (Section 9.2) are fitted down from the levels up to $P < 32$. Is it appropriate to interpret the dynamics beyond $P = 32$? Indeed, this is the point. However, before more data are available for $P > 32$, for the moment, we may just make the interpretation for the [SN] mode beyond $P = 32$ based on the Hamiltonian for $P < 32$.

We note that as HCP is close to the potential barrier leading to HPC, the level spacing will be close to zero. Then the classical frequency of the [SN] mode approaches zero. Spectroscopically, there is an extreme level. We note that as $P = 38$, the [SN] mode energy attains its maximum as shown in Fig. 9.6. The corresponding energy is 18 766.85 cm^{-1}. This value is with respect to the ground level. With respect to the potential bottom, it is 21 754.57 cm^{-1}. (The ground level with respect to the potential bottom is 2987.72 cm^{-1}). This value is close to but yet smaller than the reported barrier from HCP to HPC.[2] If the effective energy on the bending motion is considered, this value reduces to 17 640.84 cm^{-1}. (With respective to the potential bottom, it is 20 628.56 cm^{-1}.) This is short of 6 300 cm^{-1} to overcome the barrier to HPC. Hence, for $P = 38$, the $[SN]$ mode energy is not enough for the HPC formation. For P larger than 38, the $[SN]$ mode energy decreases instead. Hence, we recognize that due to nonlinear effect, the isomerization of HPC via the localized $[SN]$ mode is not realistic. The difficulty is not to overcome IVR, but that $[SN]$ mode energy is just not high enough. We found also that as n_1 is larger than 0, the nonlinear coupling will affect the bending and C–P stretching motions. However, then the $[SN]$ mode energy is decreased more. Of course, the above interpretations are based on the current Hamiltonian. It is an open question that the isomerization from HCP to HPC is possible via the excitation of the localized $[SN]$ mode. What we conclude here is that this dynamics can be more complex than expected.

9.8   The fixed point structure

Fixed points constitute the core of the dynamical structure. Fixed points not only *define* the classical properties of a dynamical system but also *define* the classification of its quantal levels and properties. For $P = 22$, Fig. 9.1 shows the locations of the four fixed points. Their $n_2/2$ and n_3 on q_2 coordinate are shown in Fig. 9.8. Their energies as P varies are shown in Fig. 9.9. (For convenience, those of $[B]$ are set to 0.) The P range for these fixed points are: for $[B]$, P is $2 - 44$; for $[r]$, P is $2 - 22$; for $[SN]$, P is $14 - 44$; for $[\overline{SN}]$, P is $14 - 22$. These data are simple. However, they are the crucial elements of a dynamical system.

9.9   DCP Hamiltonian

The system of deuterated phosphaethyne (DCP) is very similar to HCP. The deuterium replacement affects its mode frequencies such that the C–P stretching plays the role of H–C–P bending and D–C does the role of C–P stretching in HCP. The Fermi resonance in DCP is between the D–C and C–P stretching motions instead that

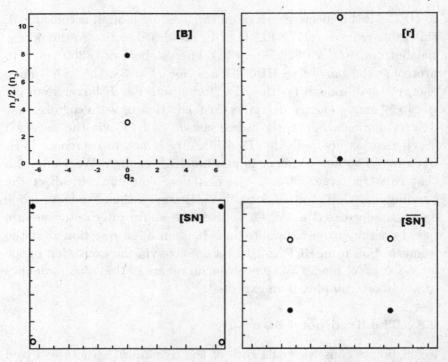

Fig. 9.8   For HCP, $P = 22$, $n_2/2$ and n_3 on q_2 coordinate for the four fixed points. Black point is $n_2/2$. Open circle is n_3. Refer to Figs. 9.1 and 9.2.

between the C–P stretching and H–C–P bending motions in HCP. Its Hamiltonian[3] is

$$H(n_1, n_2, n_3)$$

$$= \omega_1 \left(n_1 + \frac{1}{2} \right) + \omega_2 (n_2 + 1) + \omega_3 \left(n_3 + \frac{1}{2} \right)$$

$$+ X_{11} \left(n_1 + \frac{1}{2} \right)^2 + X_{12} \left(n_1 + \frac{1}{2} \right) (n_2 + 1)$$

$$+ X_{13} \left(n_1 + \frac{1}{2} \right) \left(n_3 + \frac{1}{2} \right) + X_{22} (n_2 + 1)^2$$

$$+ X_{23} (n_2 + 1) \left(n_3 + \frac{1}{2} \right) + X_{33} \left(n_3 + \frac{1}{2} \right)^2$$

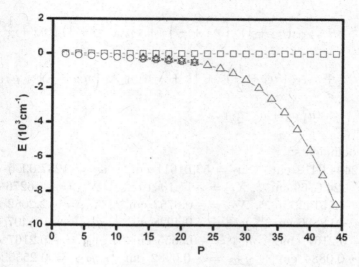

Fig. 9.9   The energies for the four fixed points. For convenience, those of $[B]$ are set to 0. $\square$: $[B]$, $\circ$: $[r]$, $\triangle$: $[SN]$, $\star$: $\overline{[SN]}$.

$$+ y_{111} \left(n_1 + \frac{1}{2} \right)^3 + y_{112} \left(n_1 + \frac{1}{2} \right)^2 (n_2 + 1)$$

$$+ y_{113} \left(n_1 + \frac{1}{2} \right)^2 \left(n_3 + \frac{1}{2} \right) + y_{122} \left(n_1 + \frac{1}{2} \right) (n_2 + 1)^2$$

$$+ y_{123} \left(n_1 + \frac{1}{2} \right) (n_2 + 1) \left(n_3 + \frac{1}{2} \right)$$

$$+ y_{133} \left(n_1 + \frac{1}{2} \right) \left(n_3 + \frac{1}{2} \right)^2 + y_{222}(n_2 + 1)^3$$

$$+ y_{223}(n_2 + 1)^2 \left(n_3 + \frac{1}{2} \right) + y_{233}(n_2 + 1) \left(n_3 + \frac{1}{2} \right)^2$$

$$+ y_{333} \left(n_3 + \frac{1}{2} \right)^3 + z_{1111} \left(n_1 + \frac{1}{2} \right)^4$$

$$+ z_{1112} \left(n_1 + \frac{1}{2} \right)^3 (n_2 + 1) + z_{1222} \left(n_1 + \frac{1}{2} \right) (n_2 + 1)^3$$

$$+ z_{1233} \left(n_1 + \frac{1}{2} \right) (n_2 + 1) \left(n_3 + \frac{1}{2} \right)^2 + z_{2222}(n_2 + 1)^4$$

$$+ z_{2233}(n_2 + 1)^2 \left(n_3 + \frac{1}{2}\right)^2 + z_{2333}(n_2 + 1)\left(n_3 + \frac{1}{2}\right)^3$$

$$+ z_{3333}\left(n_3 + \frac{1}{2}\right)^4 + \left[k + \lambda_1 n_1 + \lambda_3\left(n_3 + \frac{3}{2}\right) + \mu_{11} n_1^2\right]$$

$$\times \left(a_1^+ a_3^2 + a_3^{+2} a_1\right)$$

with coefficients:

$\omega_1 = 2494.0412$ cm^{-1}, $\omega_2 = 539.1611$ cm^{-1}, $\omega_3 = 1237.0955$ cm^{-1},
$X_{11} = -24.0769$ cm^{-1}, $X_{12} = -11.1041$ cm^{-1}, $X_{13} = -4.6276$ cm^{-1},
$X_{22} = -3.5142$ cm^{-1}, $X_{23} = -0.6753$ cm^{-1}, $X_{33} = -2.2082$ cm^{-1},
$y_{111} = -0.8896$ cm^{-1}, $y_{112} = -0.4928$ cm^{-1}, $y_{113} = -0.5407$ cm^{-1},
$y_{122} = 0.2167$ cm^{-1}, $y_{123} = -0.3655$ cm^{-1}, $y_{133} = -0.2167$ cm^{-1},
$y_{222} = 0.0884$ cm^{-1}, $y_{223} = -0.0482$ cm^{-1}, $y_{233} = 0.2535$ cm^{-1},
$y_{333} = -0.2447$ cm^{-1}, $z_{1111} = 0.0510$ cm^{-1}, $z_{1112} = 0.0806$ cm^{-1},
$z_{1222} = -0.0055$ cm^{-1}, $z_{1233} = 0.028$ cm^{-1}, $z_{2222} = -0.0014$ cm^{-1},
$z_{2233} = -0.0067$ cm^{-1}, $z_{2333} = -0.0132$ cm^{-1}, $z_{3333} = 0.0092$ cm^{-1},
$k = 12.3422$ cm^{-1}, $\lambda_1 = 0.5786$ cm^{-1}, $\lambda_3 = 0.1212$ cm^{-1}, $\mu_{11} = -0.2990$ cm^{-1}.

In this DCP system, we will be interested in the nonlinear and Fermi coupling between D–C and C–P stretchings. Hence, n_2 will be set to 0 and the conserved polyad number is $P = 2n_1 + n_3$. P will be limited to 20 due to that these coefficients were determined by fitting to the level energies with $P < 20$. The coset represented Hamiltonian is

$$H(q_3, p_3) = \omega_1\left(\frac{P}{2} - \frac{p_3^2 + q_3^2}{4} + \frac{1}{2}\right) + \omega_2 + \omega_3\left(\frac{p_3^2 + q_3^2}{2} + \frac{1}{2}\right)$$

$$+ X_{11}\left(\frac{P}{2} - \frac{p_3^2 + q_3^2}{4} + \frac{1}{2}\right)^2 + X_{12}\left(\frac{P}{2} - \frac{p_3^2 + q_3^2}{4} + \frac{1}{2}\right)$$

$$+ X_{13}\left(\frac{P}{2} - \frac{p_3^2 + q_3^2}{4} + \frac{1}{2}\right)\left(\frac{p_3^2 + q_3^2}{2} + \frac{1}{2}\right) + X_{22}$$

$$+ X_{23}\left(\frac{p_3^2 + q_3^2}{2} + \frac{1}{2}\right) + X_{33}\left(\frac{p_3^2 + q_3^2}{2} + \frac{1}{2}\right)^2$$

$$+ y_{111} \left(\frac{P}{2} - \frac{p_3^2 + q_3^2}{4} + \frac{1}{2} \right)^3 + y_{112} \left(\frac{P}{2} - \frac{p_3^2 + q_3^2}{4} + \frac{1}{2} \right)^2$$

$$+ y_{113} \left(\frac{P}{2} - \frac{p_3^2 + q_3^2}{4} + \frac{1}{2} \right)^2 \left(\frac{p_3^2 + q_3^2}{2} + \frac{1}{2} \right)$$

$$+ y_{122} \left(\frac{P}{2} - \frac{p_3^2 + q_3^2}{4} + \frac{1}{2} \right) + y_{123} \left(\frac{P}{2} - \frac{p_3^2 + q_3^2}{4} + \frac{1}{2} \right)$$

$$\times \left(\frac{p_3^2 + q_3^2}{2} + \frac{1}{2} \right) + y_{133} \left(\frac{P}{2} - \frac{p_3^2 + q_3^2}{4} + \frac{1}{2} \right)$$

$$\times \left(\frac{p_3^2 + q_3^2}{2} + \frac{1}{2} \right)^2 + y_{222} + y_{223} \left(\frac{p_3^2 + q_3^2}{2} + \frac{1}{2} \right)$$

$$+ y_{233} \left(\frac{p_3^2 + q_3^2}{2} + \frac{1}{2} \right)^2 + y_{333} \left(\frac{p_3^2 + q_3^2}{2} + \frac{1}{2} \right)^3$$

$$+ z_{1111} \left(\frac{P}{2} - \frac{p_3^2 + q_3^2}{4} + \frac{1}{2} \right)^4 + z_{1112} \left(\frac{P}{2} - \frac{p_3^2 + q_3^2}{4} + \frac{1}{2} \right)^3$$

$$+ z_{1222} \left(\frac{P}{2} - \frac{p_3^2 + q_3^2}{4} + \frac{1}{2} \right) + z_{1233} \left(\frac{P}{2} - \frac{p_3^2 + q_3^2}{4} + \frac{1}{2} \right)$$

$$\times \left(\frac{p_3^2 + q_3^2}{2} + \frac{1}{2} \right)^2 + z_{2222} + z_{2233} \left(\frac{p_3^2 + q_3^2}{2} + \frac{1}{2} \right)^2$$

$$+ z_{2333} \left(\frac{p_3^2 + q_3^2}{2} + \frac{1}{2} \right)^3 + z_{3333} \left(\frac{p_3^2 + q_3^2}{2} + \frac{1}{2} \right)^4$$

$$+ \left[k + \lambda_1 \left(\frac{P}{2} - \frac{p_3^2 + q_3^2}{4} \right) + \lambda_3 \left(\frac{p_3^2 + q_3^2}{2} + \frac{3}{2} \right) \right.$$

$$\left. + \mu_{11} \left(\frac{P}{2} - \frac{p_3^2 + q_3^2}{4} \right)^2 \right] \sqrt{\frac{P}{2} - \frac{p_3^2 + q_3^2}{4}} (q_3^2 - p_3^2)$$

and $n_3 = (q_3^2 + p_3^2)/2$, $n_1 = (P - n_3)/2$. Similarly, we can have $H(q_1, p_1)$ and the corresponding equations of motion. For short, it is shown in the Appendix.

9.10   Dynamical similarity between DCP and HCP

It is very interesting to see that the dynamical potentials of DCP show similarity with those of HCP in q_2 ($P \geq 24$) and q_3 ($P \leq 20$) coordinates. As an example, the dynamical potentials of DCP with $P = 18$ are shown in Figs. 9.10(a), (b) (The energy is with respect to the ground state which is 2387.39 cm^{-1} above the bottom of the potential well.) as contrasted with those of HCP in q_2 ($P = 24$) and q_3 ($P = 20$) coordinates (Figs. 9.10(c),(d)). It is seen that the dynamical potential of DCP in q_3 coordinate is similar to the inverse of that of HCP in q_2 coordinate and that the dynamical potential of DCP in q_1 coordinate is similar to that of HCP in q_3 coordinate

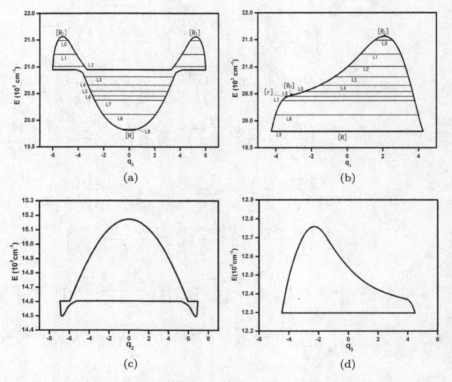

Fig. 9.10   The dynamical potentials (a), (b) of DCP with $P = 18$ as contrasted with those of HCP in q_2 ($P = 24$) (c) and q_3 ($P = 20$) coordinates (d). [R_1], [$R2$], [R] and [r] are fixed points. The horizontal lines show the quantal levels.

under $-q_3$ transformation. The comparison is made by noting that q_2 in HCP corresponds to q_3 in DCP since they both are the *slow* coordinates and that q_3 in HCP corresponds to q_1 in DCP since they both are the *fast* coordinates.

Due to this dynamical similarity, the state dynamics of DCP can be deduced properly from that of HCP. Obviously, its L0, L1 and L2 are expected to be localized with their actions mostly populated on the C–P coordinate, analogous to the $[SN]$ mode in HCP. This is shown in Fig. 9.11. These levels are also expected to be robust against IVR. From the dynamical potential in q_1 coordinate, all the levels lie in an anti-Morse type well. (At least, from L0 down to L6 as viewed from the nearest neighboring level spacings. See Fig. 9.10(b).) This is confirmed in the action integrals of these levels that the highest level possesses the least action integral and for the consecutive lower levels, the increment of action integral is a constant as shown in Table 9.2. In fact, there is deviation around L6. This peculiarity is due to that L6 is around an inflection point of the dynamical potential in q_1 coordinate. This will be addressed shortly. From the dynamical potential in q_3 coordinate, L0, L1 and L2 lie in an inverse potential while L3 to L9 lie in an harmonic-like potential. This classification of levels is confirmed by their action integrals as shown in Table 9.2. The deviation is also noted around L6. Therefore, we may classify these levels in various quantal environments by L0–L2, L3–L5, L6 and L7–L9 in the dynamical potential in q_3 coordinate and by L0–L5, L6, L7–L9 in q_1 coordinate.

For the dynamical potential of DCP in q_1 coordinate, there are four fixed points in which $[r]$ and $[R2]$ are around the inflection point as shown in Fig. 9.10(b) where L6 sits. $[r]$ is stable and $[R2]$ is unstable. (We note that in the dynamical potential of HCP in q_3 coordinate, there is an inflection point corresponding to the unstable $[\overline{SN}]$ fixed point, Fig. 9.1(b).) The other two fixed points are $[R_1]$ and $[R]$ (which is an interval in q_1 coordinate) as shown in Fig. 9.10(b). $[R_1]$ (close to L0) is mainly of the C–P stretching while $[R]$ (close to L9) is more on the D–C stretching (see Fig. 9.11). All these are consistent with Ref. [3]. L6, unlike the other levels, possesses a special character that its trajectories in (q_1, p_1) and (q_3, p_3) spaces are of

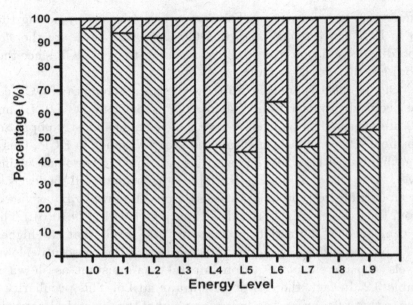

Fig. 9.11   The action percentages of $n_3/2$ (left-up shadow) and n_1 (right-up shadow) for the 10 levels of DCP with $P = 18$.

Table 9.2   The action integrals of the 10 quantal levels corresponding to $P = 18$ for DCP.

level label	Action integral in (q_1, p_1) space	Difference of the action integrals between the neighboring levels	Action integral in (q_3, p_3) space	Difference of the action integrals between the neighboring levels
L0	0.3	—	0.5	—
L1	1.3	1.0	2.5	2.0
L2	2.3	1.0	4.6	2.0
L3	3.3	1.0	11.4	—
L4	4.3	1.0	9.4	-2.0
L5	5.2	0.9	7.6	-1.8
L6	6.4	1.2	5.3	-2.3
L7	7.0	0.6	4.1	-1.2
L8	8.0	1.0	2.0	-2.1
L9	9.0	1.0	0.0	-2.0

two disconnected components. One component corresponds to more action on the C–P stretching (see Fig. 9.11). This is what addressed in Ref. [3] that L6, unlike L5 and L7, is localized on the C–P stretching viewed simply from the shape of its wave function.

In this two-mode nonlinear system with Fermi resonance, the slow mode (HCP bending and C–P stretching of DCP) shows complicated dynamics as evidenced by its dynamical potential. For the fast mode (C–P stretching of HCP and D–C stretching of DCP), the dynamical potential is more or less like an anti-Morse well. It is noted that the *peak* of the dynamical potential of HCP in q_3 coordinate is positioned in the negative region while that of the dynamical potential of DCP in q_1 coordinate is in the positive region. This implies that D–C stretching of DCP, unlike that of C–P of HCP, is more prone to dissociation since positive q coordinate corresponds to larger bond stretching displacement.[4] (See Section 1.2).

9.11   N$_2$O dynamics

As the vibrational excitation energy of N$_2$O (nitrous oxide) is not too high, the dynamics is dominated by the anharmonicity and Fermi coupling between the symmetric stretching and bending coordinates. Its algebraic Hamiltonian[5] is

$$H = H_0 + H_0^{anh} + H_1$$

$$H_0 = \omega_1 \left(n_1 + \frac{1}{2} \right) + \omega_2 (n_2 + 1) + \omega_3 \left(n_3 + \frac{1}{2} \right)$$

$$H_0^{anh} = x_{11} \left(n_1 + \frac{1}{2} \right)^2 + x_{12} \left(n_1 + \frac{1}{2} \right) (n_2 + 1)$$

$$+ x_{13} \left(n_1 + \frac{1}{2} \right) \left(n_3 + \frac{1}{2} \right) + x_{22} (n_2 + 1)^2$$

$$+ x_{23} (n_2 + 1) \left(n_3 + \frac{1}{2} \right) + x_{33} \left(n_3 + \frac{1}{2} \right)^2$$

$$+ y_{111} \left(n_1 + \frac{1}{2} \right)^3 + y_{112} \left(n_1 + \frac{1}{2} \right)^2 (n_2 + 1)$$

$$+ y_{113} \left(n_1 + \frac{1}{2} \right)^2 \left(n_3 + \frac{1}{2} \right) + y_{122} \left(n_1 + \frac{1}{2} \right) (n_2 + 1)^2$$

$$+ y_{123} \left(n_1 + \frac{1}{2} \right) (n_2 + 1) \left(n_3 + \frac{1}{2} \right) + y_{133} \left(n_1 + \frac{1}{2} \right)$$

$$\times \left(n_3 + \frac{1}{2} \right)^2 + y_{222}(n_2 + 1)^3 + y_{223}(n_2 + 1)^2 \left(n_3 + \frac{1}{2} \right)$$

$$+ y_{233}(n_2 + 1) \left(n_3 + \frac{1}{2} \right)^2 + y_{333} \left(n_3 + \frac{1}{2} \right)^3$$

$$H_1 = F_e^{(2)}(a_1^+ a_2 a_2 + h.c.).$$

The subscripts, 1, 2 and 3 are employed for symmetric stretching, N–N–O bending and antisymmetric stretching motions. The coefficients are listed in Table 9.3. n_3 is set to 0 in our treatment. The polyad number is $P = 2n_2 + n_1$. In our analysis, $P < 22$ as suggested by Ref. [5].

The coset represented Hamiltonian is:

$$H_0 = \omega_1 \left(\frac{1}{2}(p_1^2 + q_1^2) + \frac{1}{2} \right) + \omega_2(P - (p_1^2 + q_1^2) + 1) + \omega_3 \left(0 + \frac{1}{2} \right)$$

$$\dot{H}_0^{anh} = x_{11} \left(\frac{1}{2}(p_1^2 + q_1^2) + \frac{1}{2} \right)^2 + x_{12} \left(\frac{1}{2}(p_1^2 + q_1^2) + \frac{1}{2} \right)$$

Table 9.3   The coefficients for the N$_2$O algebraic Hamiltonian.

Parameter (cm^{-1})		Parameter (cm^{-1})	
ω_1	1298.59011	y_{112}	−0.116084
ω_2	596.2937	y_{113}	−0.34311
ω_3	2281.99814	y_{122}	−0.035329
X_{11}	−3.9178	y_{123}	0.51513
X_{12}	−3.0087	y_{133}	0.05979
X_{13}	−27.20721	y_{222}	−0.0131887
X_{22}	0.5432	y_{223}	0.046411
X_{23}	−14.58513	y_{233}	0.009261
X_{33}	−15.16516	y_{333}	0.015737
y_{111}	−0.004714	$F_e^{(2)}$	−17.963240

$$\times (P - (p_1^2 + q_1^2) + 1) + x_{13} \left(\frac{1}{2}(p_1^2 + q_1^2) + \frac{1}{2} \right) \left(0 + \frac{1}{2} \right)$$

$$+ x_{22}(P - (p_1^2 + q_1^2) + 1)^2 + x_{23}(P - (p_1^2 + q_1^2) + 1) \left(0 + \frac{1}{2} \right)$$

$$+ x_{33} \left(0 + \frac{1}{2} \right)^2 + y_{111} \left(\frac{1}{2}(p_1^2 + q_1^2) + \frac{1}{2} \right)^3$$

$$+ y_{112} \left(\frac{1}{2}(p_1^2 + q_1^2) + \frac{1}{2} \right)^2 (P - (p_1^2 + q_1^2) + 1)$$

$$+ y_{113} \left(\frac{1}{2}(p_1^2 + q_1^2) + \frac{1}{2} \right)^2 \left(0 + \frac{1}{2} \right) + y_{122} \left(\frac{1}{2}(p_1^2 + q_1^2) + \frac{1}{2} \right)$$

$$\times (P - (p_1^2 + q_1^2) + 1)^2 + y_{123} \left(\frac{1}{2}(p_1^2 + q_1^2) + \frac{1}{2} \right)$$

$$\times (P - (p_1^2 + q_1^2) + 1) \left(0 + \frac{1}{2} \right) + y_{133} \left(\frac{1}{2}(p_1^2 + q_1^2) + \frac{1}{2} \right)$$

$$\times \left(0 + \frac{1}{2} \right)^2 + y_{222}(P - (p_1^2 + q_1^2) + 1)^3$$

$$+ y_{223}(P - (p_1^2 + q_1^2) + 1)^2 \left(0 + \frac{1}{2} \right)$$

$$+ y_{233}(P - (p_1^2 + q_1^2) + 1) \left(0 + \frac{1}{2} \right)^2 + y_{333} \left(0 + \frac{1}{2} \right)^3$$

$$H_1 = \sqrt{2} F_e^{(2)} q_1 (P - (p_1^2 + q_1^2))$$

with $n_1 = (q_1^2 + p_1^2)/2$, $n_2 = P - (q_1^2 + p_1^2)$.

Shown in Fig. 9.12 are the two dynamical potentials in terms of q_1 and q_2 coordinates for $P = 22$, in which there are 12 levels embedded. It is interesting to note that even for other smaller polyad numbers, their dynamical potentials are very similar. This shows that the dynamics does not change significantly from the very low excitation up to $P = 22$. This situation is unlike the cases of HCP, DCP, HOCl and HOBr (See next section) where the dynamics is quite dependent on the excitation energy or the polyad number. In Fig. 9.12, fixed points are also shown. They are stable. Clearly, the dynamical potentials are defined by these fixed points.

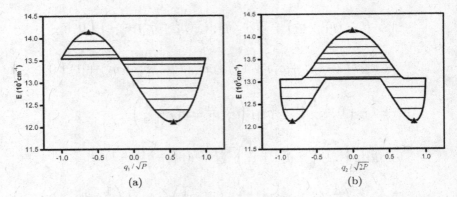

Fig. 9.12   The dynamical potentials in q_1 (a) and q_2 (b) coordinates, respectively. The level energies are with respect to the zero point energy which is $2366.8\ \mathrm{cm}^{-1}$ above the potential well bottom. The stable fixed points, as shown by ▲, are on the top and bottom of the dynamical potentials.

The action populations, (n_1, q_i) and $(n_2/2, q_i)$ can be readily obtained as shown in Figs. 9.13(a),(b). (Levels are labeled by L0, L1,... from the highest one down to the lowest one.) It is noted that except the highest and lowest levels, the curves of (n_1, q_i) and $(n_2/2, q_i)$ intersect somewhere. The intersection is especially serious for those mid-levels. The implication is that the exchange of actions (energies) on the symmetric stretching and the bending coordinates is significant. This IVR will forbid the action (energy) localization. For the highest and lowest levels, we note that the action on the bending coordinate is always larger than that on the symmetric stretching coordinate. Of course, quantum mechanically, tunneling effect may enhance IVR in these two levels. However, as will be shown below, only the lowest level is the localized mode against the energy excitation.

Shown in Fig. 9.14 are the action percentages for the 12 levels sharing $P = 22$. It is noted that the bending motion contributes more to the level dynamics, in general, except those mid-levels where IVR is strong as can be expected from the action population plots shown in Figs. 9.13(a),(b).

Shown in Fig. 9.15 are the action percentages for the lowest level, the lowest level in the upper anti-Morse well in the q_2 coordinate,

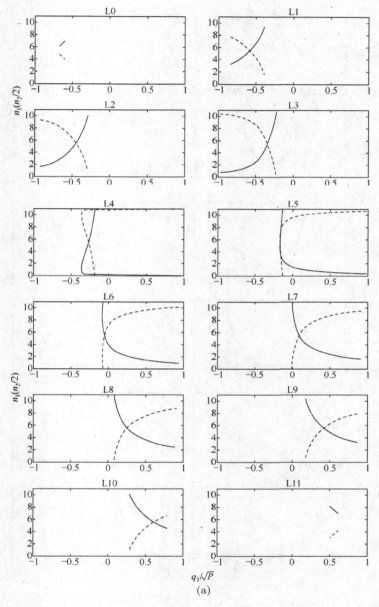

Fig. 9.13   The action populations of (n_1, q_1) (broken line) and $(n_2/2, q_1)$ (solid line) (a) and the action populations of (n_1, q_2) (broken line) and $(n_2/2, q_2)$ (solid line) (b).

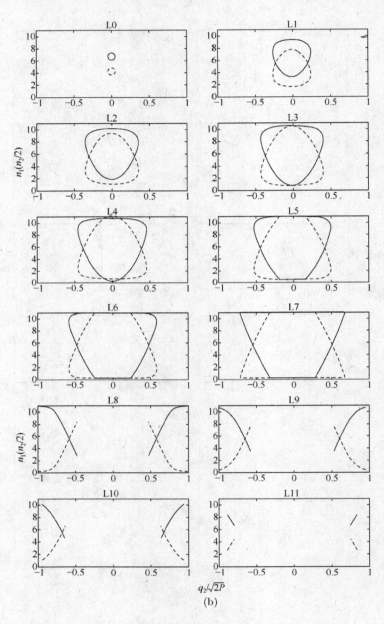

Fig. 9.13   (*Continued*)

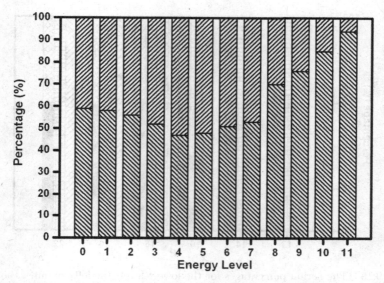

Fig. 9.14   The action percentages for the 12 levels corresponding to $P = 22$. The lower shadow is for the bending coordinate and the upper shadow is for the symmetric stretching coordinate. The levels are numbered from the highest to the lowest as L0 to L11.

(this level is close to the medium one among the levels corresponding to a given polyad number) and the highest level for the polyad number ranging from 12 to 22. It is clearly seen that the dynamical components of the symmetric stretching and bending coordinates for the lowest level in the upper anti-Morse well are not robust against the polyad number. As the polyad number increases, its action percentage of the bending coordinate decreases. Strong IVR is the main cause. For the highest level, the dynamical component of the bending coordinate is significant and IVR is not strong. However, the dynamical component is not robust against the polyad number. In this sense, we cannot categorize this level as a localized mode. Instead, the bending component of the lowest level, which can be up to more than 90%, stays persistent against the polyad number ranging from 12 to 22. So, it is a real localized bending mode. We have met with this situation in the cases of HCP and DCP. This situation will happen again in HOCl and HOBr (to be shown below). The criterion for the appearance of the localized mode is that it is born in the level

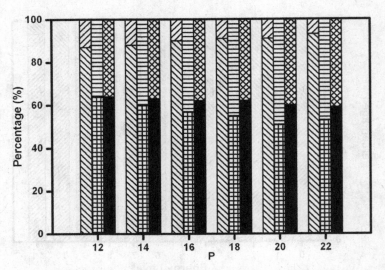

Fig. 9.15   The action percentages for the lowest level (the left column), the low-est level in the upper anti-Morse well in the q_2 coordinate (the middle column) and the highest level (the right column) for the polyad number P ranging from 12 to 22. The lower shadow is the action percentage of the bending coordinate and the upper shadow is that of the symmetric stretching coordinate.

which possesses an interval centred around nonzero q_i and lies just above a stable fixed point in the dynamical potential as was inferred in Section 9.6.

Shown in Table 9.4 are the action integrals for the levels cor-responding to $P = 22$, calculated from (q_1, p_1) and (q_2, p_2) spaces. From the action integrals calculated by (q_1, p_1), the levels can be grouped into two subsets: L0 to L3 and L4 to L11. For the former set, the action integral increases almost with a constant as the level goes down in the anti-Morse well they stay. For the latter set, the action integral decreases (also almost with a constant) as the level goes down in the Morse well they stay. From the action integrals cal-culated in (q_2, p_2) space, the levels can be grouped into two subsets, also with constant increment or decrement in their action integrals: L0 to L7 and L8 to L11. Note there is a minor deviation on L4. The constant action integral increment/decrement demonstrates the com-patibility of our classical treatment for the quantized levels. These classifications also show that the levels, though belonging to the same

Table 9.4   The action integrals for the levels corresponding to $P = 22$.

Level label	Action integral in (q_1, p_1) space	Difference between neighboring levels	Action integral in (q_2, p_2) space	Difference between neighboring levels
L0	0.03	—	0.07	—
L1	1.02	0.99	2.01	1.94
L2	2.00	0.98	4.03	2.02
L3	2.95	0.95	5.91	1.88
L4	7.19	—	7.59	1.68
L5	6.18	−1.01	9.62	2.03
L6	5.13	−1.05	11.69	2.07
L7	4.12	−1.01	13.73	2.04
L8	3.11	−1.01	6.23	—
L9	2.10	−1.01	4.21	−2.02
L10	1.08	−1.02	2.19	−2.02
L11	0.08	−1.00	0.17	−2.02

polyad number, stay in various dynamical environments which are defined essentially by the classical fixed points.

Dynamical potential method is to study the system dynamics from its levels as a whole. This is a *global* viewpoint. This is differ-ent from the method based on the wave function which emphasizes individual level and not the relation among the levels. Wave func-tion algorithm is from a *local* viewpoint. From the dynamical poten-tial, the fixed points can be easily identified. They are the essential dynamical variables. Though fixed points are classical, they *define* the quantal levels, their properties and characters. This merit enables us the way to compare various dynamical systems. This is impossible by the wave function algorithm. It is noted that different systems may still share dynamical similarity and even common properties, including the localized modes.

9.12   The cases of HOCl and HOBr

The algebraic Hamiltonians of HOCl and HOBr including their coefficients have been determined in the literatures.[6,7] For nota-tion, the subscripts, 1, 2 and 3 are used for H–O stretching,

Table 9.5   The coefficients of the algebraic Hamiltonians for HOCl and HOBr.

(cm^{-1})	HOCl	HOBr	(cm^{-1})	HOCl	HOBr	(cm^{-1})	HOCl	HOBr
ω_1	3777.067	3769.5381	y_{133}	-0.0532	-0.9150	y_{11122}	0	0.257802
ω_2	1258.914	1187.7106	y_{223}	-0.0802	-0.6071	y_{11222}	0	0.257802
ω_3	753.834	622.3722	y_{233}	0.2503	0	y_{11223}	0	0.032641
X_{11}	-80.277	-70.7877	y_{1133}	0	0.203607	y_{12222}	0	-0.020228
X_{22}	-3.204	-5.4432	y_{2222}	-0.04117	0	y_{22333}	-0.00066	-0.001709
X_{33}	-7.123	-3.5507	y_{1333}	0	0.0121098	y_{22233}	0	0.003254
X_{12}	-19.985	-26.8651	y_{3333}	0.00171	-0.000603	k	0	0
X_{13}	0	10.3315	y_{1122}	-0.15070	0	k_1	0	0.8717
X_{23}	-10.637	-3.7702	y_{1222}	0.13189	0	k_2	-0.76017	-0.3183
y_{111}	-0.3619	0	y_{2333}	-0.01229	-0.001461	k_3	-0.24939	-0.1935
y_{113}	0	-3.4216	y_{1233}	0.02381	0	k_{22}	-0.01158	-0.016988
y_{333}	0.0825	0	y_{22222}	0.00151	0	k_{23}	0.04075	0
y_{122}	-1.9534	-1.1955	y_{11112}	0	-0.656865	k_{33}	0.00583	0

H–O–Cl/H–O–Br bending and O–Cl/O–Br stretching. The coefficients are listed in Table 9.5. For these two systems, the H–O stretching basically does not couple with the bending and O–Cl/O–Br stretching motions. The latter two are coupled via Fermi resonance. The algebraic Hamiltonian is:

$$H = \omega_1 \left(n_1 + \frac{1}{2}\right) + \omega_2 \left(n_2 + \frac{1}{2}\right) + \omega_3 \left(n_3 + \frac{1}{2}\right) + X_{11} \left(n_1 + \frac{1}{2}\right)^2$$

$$+ X_{22} \left(n_2 + \frac{1}{2}\right)^2 + X_{33} \left(n_3 + \frac{1}{2}\right)^2 + X_{12} \left(n_1 + \frac{1}{2}\right)\left(n_2 + \frac{1}{2}\right)$$

$$+ X_{23} \left(n_2 + \frac{1}{2}\right)\left(n_3 + \frac{1}{2}\right) + X_{13} \left(n_1 + \frac{1}{2}\right)\left(n_3 + \frac{1}{2}\right)$$

$$+ y_{111} \left(n_1 + \frac{1}{2}\right)^3 + y_{113} \left(n_1 + \frac{1}{2}\right)^2\left(n_3 + \frac{1}{2}\right) + y_{333} \left(n_3 + \frac{1}{2}\right)^3$$

$$+ y_{122} \left(n_1 + \frac{1}{2}\right)\left(n_2 + \frac{1}{2}\right)^2 + y_{133} \left(n_1 + \frac{1}{2}\right)\left(n_3 + \frac{1}{2}\right)^2$$

$$+ y_{223} \left(n_2 + \frac{1}{2}\right)^2\left(n_3 + \frac{1}{2}\right) + y_{233} \left(n_2 + \frac{1}{2}\right)\left(n_3 + \frac{1}{2}\right)^2$$

$$+ y_{1133} \left(n_1 + \frac{1}{2}\right)^2\left(n_3 + \frac{1}{2}\right)^2 + y_{2222} \left(n_2 + \frac{1}{2}\right)^4$$

$$+ y_{1333} \left(n_1 + \frac{1}{2}\right)\left(n_3 + \frac{1}{2}\right)^3 + y_{3333} \left(n_3 + \frac{1}{2}\right)^4$$

$$+ y_{1122} \left(n_1 + \frac{1}{2}\right)^2\left(n_2 + \frac{1}{2}\right)^2 + y_{1222} \left(n_1 + \frac{1}{2}\right)\left(n_2 + \frac{1}{2}\right)^3$$

$$+ y_{2333} \left(n_2 + \frac{1}{2}\right)\left(n_3 + \frac{1}{2}\right)^3 + y_{1233} \left(n_1 + \frac{1}{2}\right)\left(n_2 + \frac{1}{2}\right)$$

$$\times \left(n_3 + \frac{1}{2}\right)^2 + y_{22222} \left(n_2 + \frac{1}{2}\right)^5 + y_{11112} \left(n_1 + \frac{1}{2}\right)^4$$

$$\times \left(n_2 + \frac{1}{2}\right) + y_{11122} \left(n_1 + \frac{1}{2}\right)^3\left(n_2 + \frac{1}{2}\right)^2 + y_{11222} \left(n_1 + \frac{1}{2}\right)^2$$

$$\times \left(n_2 + \frac{1}{2}\right)^3 + y_{11223} \left(n_1 + \frac{1}{2}\right)^2\left(n_2 + \frac{1}{2}\right)^2\left(n_3 + \frac{1}{2}\right)$$

$$+ y_{12222} \left(n_1 + \frac{1}{2} \right) \left(n_2 + \frac{1}{2} \right)^4 + y_{22333} \left(n_2 + \frac{1}{2} \right)^2 \left(n_3 + \frac{1}{2} \right)^3$$

$$+ y_{22233} \left(n_2 + \frac{1}{2} \right)^3 \left(n_3 + \frac{1}{2} \right)^2 + \left[k + k_1 \left(n_1 + \frac{1}{2} \right) + k_2 n_2 \right.$$

$$\left. + k_3 \left(n_3 + \frac{3}{2} \right) + k_{22} n_2^2 + k_{23} n_2 \left(n_3 + \frac{3}{2} \right) + k_{33} \left(n_3 + \frac{3}{2} \right)^2 \right]$$

$$\times (a_2^+ a_3^2 + a_3^{+2} a_2).$$

Here, n_1 is set to 0, $P = 2n_2 + n_3$, $P \leq 30$ (Since the coefficients are fitted down by the level energies with $P \leq 30$).

The coset represented Hamiltonian is with $n_2 = (q_2^2 + p_2^2)/2$, $n_3 = P - (q_2^2 + p_2^2)$:

$$H(q_2, p_2, P)$$

$$= \frac{1}{2} \omega_1 + \omega_2 \left(\frac{p_2^2 + q_2^2}{2} + \frac{1}{2} \right) + \omega_3 \left(P - (p_2^2 + q_2^2) + \frac{1}{2} \right)$$

$$+ \frac{1}{4} X_{11} + X_{22} \left(\frac{p_2^2 + q_2^2}{2} + \frac{1}{2} \right)^2 + X_{33} \left(P - (p_2^2 + q_2^2) + \frac{1}{2} \right)^2$$

$$+ \frac{1}{2} X_{12} \left(\frac{p_2^2 + q_2^2}{2} + \frac{1}{2} \right) + X_{23} \left(\frac{p_2^2 + q_2^2}{2} + \frac{1}{2} \right)$$

$$\times \left(P - (p_2^2 + q_2^2) + \frac{1}{2} \right) + \frac{1}{2} X_{13} \left(P - (p_2^2 + q_2^2) + \frac{1}{2} \right) + \frac{1}{8} y_{111}$$

$$+ \frac{1}{4} y_{113} \left(P - (p_2^2 + q_2^2) + \frac{1}{2} \right) + y_{333} \left(P - (p_2^2 + q_2^2) + \frac{1}{2} \right)^3$$

$$+ \frac{1}{2} y_{122} \left(\frac{p_2^2 + q_2^2}{2} + \frac{1}{2} \right)^2 + \frac{1}{2} y_{133} \left(P - (p_2^2 + q_2^2) + \frac{1}{2} \right)^2$$

$$+ y_{223} \left(\frac{p_2^2 + q_2^2}{2} + \frac{1}{2} \right)^2 \left(P - (p_2^2 + q_2^2) + \frac{1}{2} \right)$$

$$+ y_{233} \left(\frac{p_2^2 + q_2^2}{2} + \frac{1}{2} \right) \left(P - (p_2^2 + q_2^2) + \frac{1}{2} \right)^2$$

$$+ \frac{1}{4} y_{1133} \left(P - (p_2^2 + q_2^2) + \frac{1}{2} \right)^2 + y_{2222} \left(\frac{p_2^2 + q_2^2}{2} + \frac{1}{2} \right)^4$$

$$+ \frac{1}{2} y_{1333} \left(P - (p_2^2 + q_2^2) + \frac{1}{2} \right)^3 + y_{3333} \left(P - (p_2^2 + q_2^2) + \frac{1}{2} \right)^4$$

$$+ \frac{1}{4} y_{1122} \left(\frac{p_2^2 + q_2^2}{2} + \frac{1}{2} \right)^2 + \frac{1}{2} y_{1222} \left(\frac{p_2^2 + q_2^2}{2} + \frac{1}{2} \right)^3$$

$$+ y_{2333} \left(\frac{p_2^2 + q_2^2}{2} + \frac{1}{2} \right) \left(P - (p_2^2 + q_2^2) + \frac{1}{2} \right)^3$$

$$+ \frac{1}{2} y_{1233} \left(\frac{p_2^2 + q_2^2}{2} + \frac{1}{2} \right) \left(P - (p_2^2 + q_2^2) + \frac{1}{2} \right)^2$$

$$+ y_{22222} \left(\frac{p_2^2 + q_2^2}{2} + \frac{1}{2} \right)^5 + \frac{1}{16} y_{11112} \left(\frac{p_2^2 + q_2^2}{2} + \frac{1}{2} \right)$$

$$+ \frac{1}{8} y_{11122} \left(\frac{p_2^2 + q_2^2}{2} + \frac{1}{2} \right)^2 + \frac{1}{4} y_{11222} \left(\frac{p_2^2 + q_2^2}{2} + \frac{1}{2} \right)^3$$

$$+ \frac{1}{4} y_{11223} \left(\frac{p_2^2 + q_2^2}{2} + \frac{1}{2} \right)^2 \left(P - (p_2^2 + q_2^2) + \frac{1}{2} \right)$$

$$+ \frac{1}{2} y_{12222} \left(\frac{p_2^2 + q_2^2}{2} + \frac{1}{2} \right)^4 + y_{22333} \left(\frac{p_2^2 + q_2^2}{2} + \frac{1}{2} \right)^2$$

$$\times \left(P - (p_2^2 + q_2^2) + \frac{1}{2} \right)^3 + y_{22233} \left(\frac{p_2^2 + q_2^2}{2} + \frac{1}{2} \right)^3$$

$$\times \left(P - (p_2^2 + q_2^2) + \frac{1}{2} \right)^2 + \left[k + \frac{1}{2} k_1 + k_2 \frac{p_2^2 + q_2^2}{2} \right.$$

$$+ k_3 \left(P - (p_2^2 + q_2^2) + \frac{3}{2} \right) + k_{22} \left(\frac{p_2^2 + q_2^2}{2} \right)^2 + k_{23} \frac{p_2^2 + q_2^2}{2}$$

$$\times \left. \left(P - (p_2^2 + q_2^2) + \frac{3}{2} \right) + k_{33} \left(n_3 + \frac{3}{2} \right)^2 \right] \sqrt{2} (P - (p_2^2 + q_2^2)) q_2.$$

The dynamical potential, fixed point, action population, action percentage and action integral are the tools we will employ for the dynamical analysis. The dynamical potentials are mostly composed of two parts: the upper one is an anti-Morse well and the lower one is a (deformed) Morse well. The fixed points which possess bending character are shown by B. If they are of more O–Cl/O–Br stretching character, the notation R is employed. An upper bar is added for the unstable fixed points, otherwise, they are the stable one. Solid star/open star denote the second stable/unstable B, R. 2 and 3 are for reminding that fixed points are derived from the potentials in q_2 or q_3 coordinate.

In the followings, the dynamical properties of HOCl are shown.

The levels of the HOCl system are shown in Fig. 9.16. These levels are grouped by the polyad number P. Shown therein are also the fixed points appearing in the corresponding dynamical potentials (shown below). By the fixed points, the levels are grouped into five categories for P from 10 to 30. They are 10–18, 20, 22, 24 and 26–30. We note that the dynamical structure changes most as the polyad number is around 20, 22 and 24. For the levels below and above these energy realms, the dynamical structure remains quite unchanged. Shown in Fig. 9.17 are the dynamical potentials in q_2 and q_3 coordinates with P being 12, 20, 22, 24 and 30 for showing the dynamical structural characters of these five categories. The horizontal lines shown therein are the quantized levels. We note that quantized levels sharing a common polyad number are enclosed in their respective dynamical potential. Also shown are the fixed points. The lower portion of the dynamical potential is of (deformed and even multiple) Morse type while the upper portion is of an (deformed and even multiple) anti-Morse type. What impressive is that the nearest level spacing around the top of the Morse well or the bottom of the anti-Morse well is usually narrower. These regions correspond to the dynamical separatrix and often therefrom the unstable fixed points are born. This is most evident in the cases as P is 24 and 30. This narrowing of the nearest level spacing would be difficult to understand without the dynamical potential. We further note that the dynamical potentials with $P = 24$ are very similar to those of HCP with $P = 22$ though the

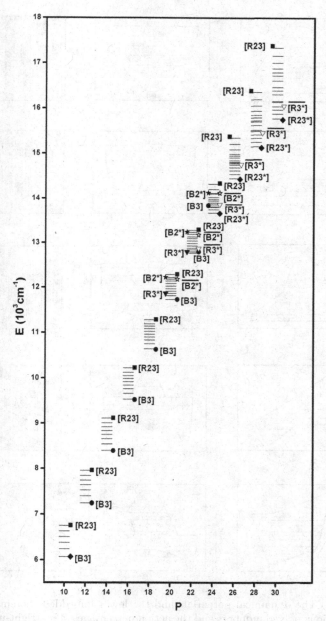

Fig. 9.16   The levels for various P up to 30 for HOCl. The level energies are with respect to the zero point energies which are 2864.2 cm^{-1} above the potential well bottoms.

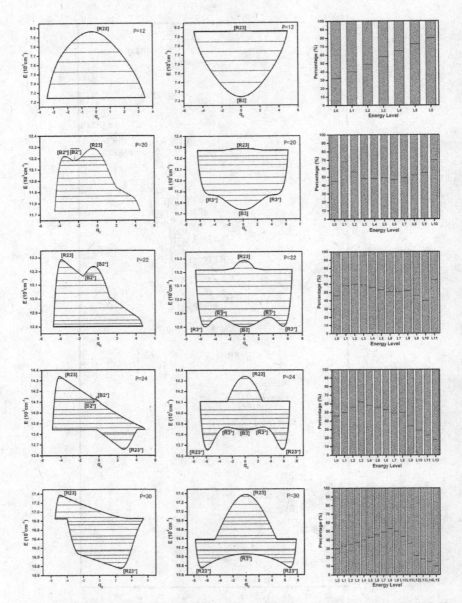

Fig. 9.17   The dynamical potentials and the levels imbedded therein for HOCl under various polyad numbers and the action percentages of n_2 (right-up shadow) and $n_3/2$ (left-up shadow) for the respective levels. Note that the levels are numbered from the highest to the lowest as L0 to L($P/2$) for each P. Also shown are the fixed points.

q_2 and q_3 coordinates are swapped. The property of bending localization in HCP is therefore reflected in the localization of the O–Cl stretching mode in this HOCl system.

Shown in Fig. 9.17 are also the action percentages for the corresponding levels. The levels are numbered from the highest to the lowest as L0 to L($P/2$) for each P. For small P, the higher levels possess more actions in the O–Cl stretching coordinate and the lower levels possess more actions in the H–O–Cl bending coordinate. As P reaches 24, an abrupt change occurs that the very lowest level possesses localized actions in the O–Cl coordinate. As mentioned, this localization of action is observed in the HCP and N₂O bending coordinate. We note that these cases of action localization share a common character in their dynamical potentials as mentioned in Section 9.6. The character is that: localization is born in the level which possesses an interval centered around nonzero q_i and lies just above a stable fixed point in the dynamical potential. This localization is further demonstrated in the action population analysis.

Shown in Fig. 9.18 are n_2 and $n_3/2$ as a function of q_2 for the levels with $P = 30$, for demonstration. (Those as a function of q_3 show identical conclusions.) The localization of actions is observed in the lowest four levels, L12–L15 for which $n_3/2$ is always larger than n_2. For the higher levels L0 to L4, though $n_3/2$ is also larger than n_2, the action interchange between the q_2 and q_3 coordinates is eminently more violent than that in L12–L15. For the levels in between, IVR between the bending and O–Cl stretching is rather complete. (See the cross-over of the n_2 and $n_3/2$ curves.) These level characters are identical to those by the quantal wave functions, except that the quantal calculation indicates that L10, L12[8] (or L11, L13[8]) are drastically more of bending character, unlike their neighboring levels. In fact, this discrepancy is not fatal. We note that these levels are very close to the unstable fixed point $[\overline{R3^*}]$ so that if the motion stays on it ($q_3 = 0$), the action (energy) will be mostly localized on the bending coordinate just as evidenced by the quantal calculation. This *discrepancy* is found in the levels close to the unstable fixed points, in general. This will be also found in the HOBr case. In fact,

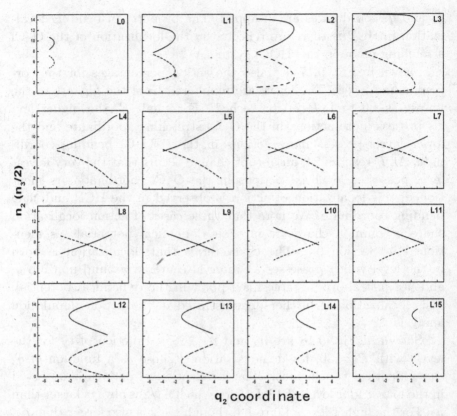

Fig. 9.18 The action populations as a function of q_2 coordinate for the levels with $P = 30$ for HOCl. Solid line is for $n_3/2$ and the broken line is for n_2.

this is the *scarring effect* by Heller[9] that there is enhancement of the wave function amplitude around the unstable periodic orbits.

The action integrals for the levels sharing a common polyad number are useful in categorizing the level character. The case for $P = 30$ is shown in Table 9.6. Indeed, we note that for the levels from the highest one down in the upper region of the dynamical potential either in q_2 or q_3 coordinate, the action integral increases with a constant increment. This corresponds to the character of an anti-Morse type potential. While for the lower levels from the lowest one up, the action integral increases with a constant increment, since the corresponding potential is a Morse type. We note that the

Table 9.6   The action integrals of the 16 quantal levels corresponding to $P = 30$ for HOCl.

Level label	Action integral in (q_2, p_2) space	Difference of the action integrals between the neighboring levels	Action integral in (q_3, p_3) space	Difference of the action integrals between the neighboring levels
L0	0.3	—	0.3	—
L1	1.4	1.1	2.0	1.7
L2	2.5	1.1	3.9	1.9
L3	3.5	1.0	5.9	2
L4	11.1	—	7.7	1.8
L5	10.2	−0.9	9.6	1.9
L6	9.3	−0.9	11.5	1.9
L7	8.3	−1.0	13.3	1.8
L8	7.4	−0.9	16.0	—
L9	6.3	−1.1	14.0	−2.0
L10	5.3	−1.0	11.8	−2.2
L11	4.2	−1.1	9.4	−2.4
L12	3.4	−0.8	7.2	−2.2
L13	2.4	−1.0	4.9	−2.3
L14	1.2	−1.2	2.6	−2.3
L15	0.2	−1.0	0.3	−2.3

interpretation of the level property from the dynamical potential and that by the action integral are two independent approaches, while both lead to a consistent conclusion.

We note that action localization only occurs in the O–Cl bond and not on the bending motion. This is contrary to the HCP case. Therefore, we access that the isomerization to HClO through the internal rotation of H atom is difficult. Instead, the dissociation of O–Cl bond is possible from the viewpoint of action (energy) localization.

The dynamical properties of HOBr are shown below.

The previous study on the HOCl system offers us a routine analysis for the HOBr case. Shown in Fig. 9.19 is the energy diagram for $P = 10$ up to 30 together with the fixed points which are more obvious from the dynamical potentials as shown in Fig. 9.20. From the fixed point structure, the vibrational dynamics can be grouped into two categories: $P = 10$ to 14, 16 to 30. (The dynamical potential

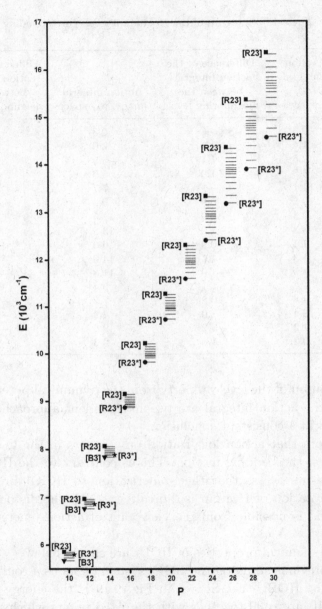

Fig. 9.19   The levels for P up to 30 for HOBr. The level energies are with respect to the zero point energies which is $2764.0\,\mathrm{cm}^{-1}$ above the potential well bottom. Also shown are the fixed points.

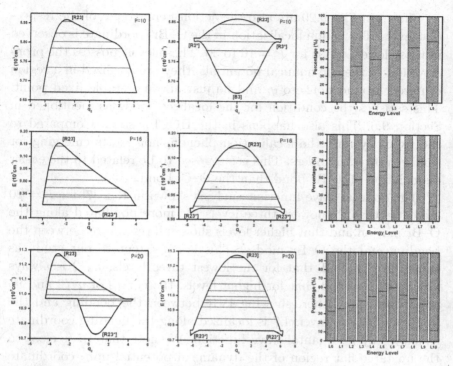

Fig. 9.20   The dynamical potentials and the levels imbedded therein for HOBr under various polyad numbers and the action percentages of n_2 (right-up shadow) and $n_3/2$ (left-up shadow) for the respective levels. Note that the levels are numbered from the highest to the lowest as L0 to L($P/2$) for each P. Also shown are the fixed points.

in q_2 coordinate for $P = 16$ shows its uniqueness and is also shown in Fig. 9.20.)

The general structure of the dynamical potentials is composed of an anti-Morse well in the upper energy realm and one or two Morse wells in the lower energy realm. The nearest neighboring energy spacing is the smallest around where the anti-Morse well and Morse well join together. This corresponds to the dynamical separatrix. The fixed point structure of HOBr system shows that it possesses much simpler dynamics than the HOCl system. Also noted is that bending character disappears for levels with larger P. This was also confirmed in Ref. [7].

Shown in Fig. 9.20 are also the action percentages of the respective levels. The action localization in the O–Br coordinate is observed in those lowest levels for $P = 16$ to 30. These levels possess the property that in their dynamical potentials, they are localized in a region centered around nonzero q_i and lie just above a stable fixed point. This satisfies the condition for the localized mode as mentioned in Section 9.6. This also happens in the HOCl system. Compared to the HOCl system, the localization phenomenon starts emerging for smaller polyad numbers. This is supposed to be related to the easier dissociation of O–Br bond than the O–Cl bond.

The quantal wave function analysis shown in Ref. [7] for $P = 20$ indicates that the lowest three levels are more populated along the O–Br motion and that higher levels show full resonance between the bending and the O–Br modes. This is consistent to our study as shown in Fig. 9.21 that for the lowest three levels, $n_3/2$ is always larger than n_2 and that for higher levels, the curves of $n_3/2$ and n_2 cross over each other, showing IVR between the bending and the O–Br modes. However, L7 is localized along the bending coordinate as shown by its quantal wave function. We note that L7 is close to the unstable flat region of the dynamical potential in q_2 coordinate as shown in Fig. 9.20. Therefore, for this level, if the motion stays fixed on large q_2 coordinate, it will be mainly along the bending coordinate. So, this subtlety predicted by the quantal algorithm can be reasoned by this classical dynamical analysis that this level is near the unstable region.

The action integrals for the levels with $P = 20$ are shown in Table 9.7. In the (q_2, p_2) space, for the higher levels in the upper anti-Morse well, the action integral increases from the top level down with a constant increment. For the lower levels, the action integral increases from the lowest level up with a constant increment, showing a Morse well character. In the (q_3, p_3) space, except the lowest level, all the level action integrals show the behavior of an anti-Morse well. The consistent implications by the action integral and the dynamical potential are well established.

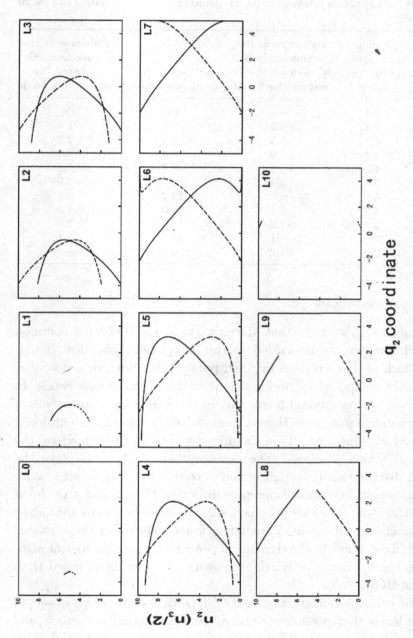

Fig. 9.21 The action populations as a function of q_2 coordinate for the levels with $P = 20$ for HOBr. Solid line is for $n_3/2$ and the broken line is for n_2.

Table 9.7 The action integrals of the 11 quantal levels corresponding to $P = 20$ for HOBr.

Level label	Action integral in (q_2, p_2) space	Difference of the action integrals between the neighboring levels	Action integral in (q_3, p_3) space	Difference of the action integrals between the neighboring levels
L0	0.3	—	0.3	—
L1	1.2	0.9	2.2	1.9
L2	2.3	1.1	4.3	2.1
L3	3.3	1.0	6.3	2.0
L4	4.3	1.0	8.1	1.8
L5	5.5	1.2	10.1	2.0
L6	5.0	—	12.2	2.1
L7	3.9	−1.1	14.0	1.8
L8	2.6	−1.3	16.0	2.0
L9	1.5	−1.1	18.1	2.1
L10	0.3	−1.2	0.3	—

9.13 A comment

In conclusion, we note that the quantal levels sharing a common polyad number are imbedded in the associated dynamical potential which is characterized by fixed points. The dynamical potential is mainly composed of (deformed) anti-Morse and Morse wells. In other words, the quantal levels stay in the multiple environments of the dynamical potential. Hence, these environments can be employed for level classification. This classification can be confirmed by the classical action integrals. The classical analysis also employs the action distribution (population and percentage). This classical analysis can lead to the most inferences drawn by the quantal wave function algorithm. The subtle behavior of the wave functions that show drastic change of mode character is found related to the unstable region/fixed point in the dynamical potential. By this classical algorithm, the dynamics of HOBr is shown to be less complicated than that of HOCl.

The action localization in the O–Cl/O–Br coordinates is observed in the levels that possess an interval centred around nonzero q_i and lie just above a stable fixed point in the dynamical potential. This is true for the HCP, DCP, N_2O cases. This is a general rule for the action localization as stated in Section 9.6. The localization analysis demonstrates that O–Br bond is more prone to dissociation than O–Cl bond.

This approach is global in the sense that the focus is on a set of levels instead on the individual one. The significance of this work is that the vibrational dynamics can be unfolded out by the classical nonlinear variables. Quantal levels are, in fact, related to the classical dynamical properties/variables signified by the fixed point structure. Fixed point structure, though classical, plays the core role in a quantal system. The global dynamical approach also enables us to recognize the dynamical similarity among the various systems. This is impossible or very difficult by the wave function algorithm.

References

1. Joyeux M, Sugny D, Tyng V, Kellman M, Ishikawa H, Field R, Beck C, Schinke R. *J. Chem. Phys.*, 2000, 112: 4162.
2. Farantos S C, Keller H M, Schinke R, Yamashita K, Morokuma K. *J. Chem. Phys.*, 1996, 104: 10055.
3. Bredenbeck J, Beck C, Schinke R, Koput J, Stamatiadis S, Farantos S C, Joyeux M. *J. Chem. Phys.*, 2000, 112: 8855.
4. Rankin C C, Miller W H. *J. Chem. Phys.*, 1971, 55: 3150.
5. Waalkens H, Jung C, Taylor H S. *J. Phys. Chem.*, 2002, A106: 911.
6. Jost R, Joyeux M, Skokov, Bowman S J. *J. Chem. Phys.*, 1999, 111: 6807.
7. Azzam T, Schinke R, Farantos S C, Joyeux M, Peterson K A. *J. Chem. Phys.*, 2003, 118: 9643.
8. Weiss J, Hauschildt J, Grebenshchikov S Y, Duren R, Schinke R, Koput J, Stamatiadis S, Farantos S C. *J. Chem. Phys.*, 2000, 112: 77.
9. Heller E J. *Phys. Rev. Lett.*, 1984, 53: 1515.

Appendix

The coset represented Hamiltonians of HCP in (q_3, p_3) (a) and DCP in (q_1, p_1) (b)

(a)

$$H(q_3, p_3) = \frac{1}{2}\omega_1 + \omega_2(P - p_3^2 - q_3^2 + 1) + \omega_3\left(\frac{p_3^2 + q_3^2}{2} + \frac{1}{2}\right)$$

$$+ \frac{1}{4}X_{11} + \frac{1}{2}X_{12}(P - p_3^2 - q_3^2 + 1) + \frac{1}{2}X_{13}\left(\frac{p_3^2 + q_3^2}{2} + \frac{1}{2}\right)$$

$$+ X_{22}(P - p_3^2 - q_3^2 + 1)^2 + X_{23}(P - p_3^2 - q_3^2 + 1)$$

$$\times \left(\frac{p_3^2 + q_3^2}{2} + \frac{1}{2} \right) + X_{33} \left(\frac{p_3^2 + q_3^2}{2} + \frac{1}{2} \right)^2$$

$$+ y_{222}(P - p_3^2 - q_3^2 + 1)^3 + z_{2222}(P - p_3^2 - q_3^2 + 1)^4$$

$$- \left[k + \frac{1}{2}k_1 + k_2(P - p_3^2 - q_3^2 + 2) + k_3 \left(\frac{p_3^2 + q_3^2}{2} \right) \right]$$

$$\times \sqrt{2}(P - p_3^2 - q_3^2)q_3.$$

(b)

$$H(q_1, p_1) = \omega_1 \left(\frac{p_1^2 + q_1^2}{2} + \frac{1}{2} \right) + \omega_2 + \omega_3 \left(P - (p_1^2 + q_1^2) + \frac{1}{2} \right)$$

$$+ X_{11} \left(\frac{p_1^2 + q_1^2}{2} + \frac{1}{2} \right)^2 + X_{12} \left(\frac{p_1^2 + q_1^2}{2} + \frac{1}{2} \right)$$

$$+ X_{13} \left(\frac{p_1^2 + q_1^2}{2} + \frac{1}{2} \right) \left(P - (p_1^2 + q_1^2) + \frac{1}{2} \right) + X_{22}$$

$$+ X_{23} \left(P - (p_1^2 + q_1^2) + \frac{1}{2} \right) + X_{33} \left(P - (p_1^2 + q_1^2) + \frac{1}{2} \right)^2$$

$$+ y_{111} \left(\frac{p_1^2 + q_1^2}{2} + \frac{1}{2} \right)^3 + y_{112} \left(\frac{p_1^2 + q_1^2}{2} + \frac{1}{2} \right)^2$$

$$+ y_{113} \left(\frac{p_1^2 + q_1^2}{2} + \frac{1}{2} \right)^2 \left(P - (p_1^2 + q_1^2) + \frac{1}{2} \right)$$

$$+ y_{122} \left(\frac{p_1^2 + q_1^2}{2} + \frac{1}{2} \right) + y_{123} \left(\frac{p_1^2 + q_1^2}{2} + \frac{1}{2} \right)$$

$$\times \left(P - (p_1^2 + q_1^2) + \frac{1}{2} \right) + y_{133} \left(\frac{p_1^2 + q_1^2}{2} + \frac{1}{2} \right)$$

$$\times \left(P - (p_1^2 + q_1^2) + \frac{1}{2} \right)^2 + y_{222}$$

$$+ y_{223} \left(P - (p_1^2 + q_1^2) + \frac{1}{2} \right) + y_{233} \left(P - (p_1^2 + q_1^2) + \frac{1}{2} \right)^2$$

$$+ y_{333} \left(P - (p_1^2 + q_1^2) + \frac{1}{2} \right)^3$$

$$+ z_{1111} \left(\frac{p_1^2 + q_1^2}{2} + \frac{1}{2} \right)^4 + z_{1112} \left(\frac{p_1^2 + q_1^2}{2} + \frac{1}{2} \right)^3$$

$$+ z_{1222} \left(\frac{p_1^2 + q_1^2}{2} + \frac{1}{2} \right) + z_{1233} \left(\frac{p_1^2 + q_1^2}{2} + \frac{1}{2} \right)$$

$$\times \left(P - (p_1^2 + q_1^2) + \frac{1}{2} \right)^2 + z_{2222}$$

$$+ z_{2233} \left(P - (p_1^2 + q_1^2) + \frac{1}{2} \right)^2$$

$$+ z_{2333} \left(P - (p_1^2 + q_1^2) + \frac{1}{2} \right)^3$$

$$+ z_{3333} \left(P - (p_1^2 + q_1^2) + \frac{1}{2} \right)^4$$

$$+ \left[k + \lambda_1 \left(\frac{p_1^2 + q_1^2}{2} \right) + \lambda_3 \left(P - (p_1^2 + q_1^2) + \frac{3}{2} \right) \right.$$

$$+ \left. \mu_{11} \left(\frac{p_1^2 + q_1^2}{2} \right)^2 \right] \sqrt{2} (P - (p_1^2 + q_1^2)) q_1.$$

Chapter 10

Chaos in the Transition State Induced by the Bending Motion

10.1　Chaos in the transition state

The classical frequency ω_c can be written as:

$$\partial E / \partial n$$

where E is energy and n is action. In this formula, if $\partial n = 1$, then ∂E is the energy spacing between two neighboring levels. Hence, the energy spacing between two neighboring levels corresponds to the classical frequency ω_c. The basic characteristic of a nonlinear system is that its classical frequency depends on its action. For a Morse oscillator, ω_c is smaller as n is larger. At dissociation, ω_c is close to 0.

The above physics is very similar to the pendulum motion. At low excitation, the motion is around the stable fixed point with small amplitude. At high excitation close to dissociation, the motion is just like that of the pendulum around the unstable fixed point, In Chapter 1, we have addressed this concept and pointed out that chaos appears around the unstable fixed point. Therefore, we conjecture that *dissociation, or more generally, a transition state is accompanied by chaos.* The transition state is an unstable delocalized state. Its classical frequency is close to zero.

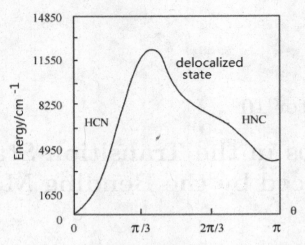

Fig. 10.1   The potential of HCN, HNC and the transition state (delocalized state). θ is the angle of H around C–N. $\theta = 0$ is HCN, $\theta = \pi$ is HNC.

Similar phenomenon appears in the high excitation of bending motion in intramolecular rotation. For example, in acetylene, the H atom of the C–H bending at high excitation will overcome the potential barrier to migrate to another C atom to form vinylidene. Barrier top corresponds to the unstable fixed point of the pendulum motion. In HCN, the highly excited bending motion of C–H may cause the H atom to overcome the energy barrier and migrate to the N atom to form the stable HNC. Figure 10.1 shows their potential.[1,2] Another example is the migration of the H atom to form linear HPC by the high excitation of the C–H bending motion in HCP. HPC is unstable whose H atom is at the unstable saddle (fixed) point. At even higher excitation, the H atom will rotate around the C–P core.

More specifically, take the HCN case as an example. The main coupling is between the C–H stretching and C–H bending motions[1] The stretching frequency ω_s is 3311 cm^{-1} and for the bending, ω_b is 713 cm^{-1}. (The coupling between the C–H bending and C–N stretching is very small.) By energy consideration, the coupling orders are 1:4 (energy difference is 459 cm^{-1}) and 1:5 (energy difference is 254 cm^{-1}). In HNC, the frequency of the N–H stretching, ω_s is 3653 cm^{-1} and that of the bending, ω_b is 477 cm^{-1}. Hence, the couplings can be of the orders 1:7 (energy difference is 314 cm^{-1}) and 1:8 (energy difference is 163 cm^{-1}). In the transition state, ω_c of the H bending around C–N will be less than 400 cm^{-1}. Then, the bending will be in multiple resonances with the stretching which is between 3311 cm^{-1} and 3653 cm^{-1}. (The stretching is at low excitation with very small nonlinear effect.) The case of HCP is similar.[3] ω_b of the C–H bending is 698 cm^{-1} which can be in 1:2 resonance with the C–P stretching whose ω_s is 1301 cm^{-1}. As the energy of the C–H bending is high enough that the system is of the linear saddle structure, CPH, the bending frequency will be reduced so that its multiple resonances with the C–P stretching are possible. Hence, we note that the transition state is accompanied by the multiple resonances which is due to that the frequency of the bending motion is then greatly reduced. Chaos in the transition state is therefore related to Chirikov's point (Section 3.4) on the emergence of chaos by the resonance overlapping.

10.2   The cases of HCN, HNC and the transition state

The quantal levels of HCN, HNC, which are stable isomers, and the transition state, which is delocalized, are available from the literature.[1,2] They are based on an *ab initio* quantal calculation. Hereby, subscripts 1, 2 and 3 are referred to the H–C stretching, the bending and the C–N stretching coordinates, respectively. The algebraic Hamiltonian contains the following Morse terms:

$$\omega_1(n_1 + 1/2) + X_{11}(n_1 + 1/2)^2 + \omega_2(n_2 + 1) + X_{22}(n_2 + 1)^2$$
$$+ \omega_3(n_3 + 1/2) + X_{33}(n_3 + 1/2)^2$$

the anharmonic terms

$$X_{12}(n_1 + 1/2)(n_2 + 1) + X_{13}(n_1 + 1/2)(n_3 + 1/2) + X_{23}(n_2 + 1)(n_3 + 1/2)$$

and the inter-mode coupling terms

$$\sum K_{k_1 k_2 k_3} O_1^{|k_1|} O_2^{|k_2|} O_3^{|k_3|}$$

where ω, X and K are coefficients in unit of cm^{-1}. n is the action. k_i's are integers. The operator O_i is taken as a^+ when $k_i > 0$ and as

Table 10.1 The coefficients in cm^{-1} of the algebraic Hamiltonians of HCN, HNC and the transition state. The energy ranges (in cm^{-1}) are with respect to the minimum of the HCN potential.

	HCN		HNC		Transition State
	4194–15809	15845–26528	8504–16184	16189–18986	18687–21827
ω_1	3442.1	3494.6	3284.1	3317.1	3342.22
ω_2	728.2	736.4	492.2	498.0	258.35
ω_3	2129.9	2109.4	2069.4	2046.0	2227.00
X_{11}	−51.4	−60.8	−69.4	−86.8	−91.66
X_{22}	−2.6	−3.7	−26.0	−5.3	0.11
X_{33}	−10.7	−7.3	−28.7	−29.1	−5.32
X_{12}	−19.1	−25.2	18.1	−32.7	−23.19
X_{13}	−15.9	−15.3	−93.5	−84.7	−27.41
X_{23}	−3.1	−3.7	−62.4	−36.9	−23.59
K_{03-1}			−9.3	−6.3	
K_{04-1}		−2.1		−5.8	
K_{10-2}	−2.9		−58.2	−46.8	
K_{11-2}		9.1	21.7	8.4	
K_{1-2-1}		3.2	−76.5	−73.9	−0.45
K_{1-3-1}		−5.6	13.6	8.7	−0.79
K_{1-4-1}					−0.13
K_{12-2}					1.09
K_{13-2}					0.17
K_{14-2}					−0.04

a when $k_i < 0$. a^+ and a are the creation and destruction operators, respectively. We expect that the resonance condition, $k_1\omega_1 + k_2\omega_2 + k_3\omega_3$, is not too large, so is $|k_1| + |k_2| + |k_3|$ which is the resonance order.

By this algebraic Hamiltonian, a Hamiltonian matrix can be constructed via the basis states $|n_1\rangle|n_2\rangle|n_3\rangle$. (For the transition state, n_1 and n_3 are 0, 1, 2, n_2 is 40 to 70. This is because that the transition state is mainly due to the excitation of the bending motion.) The eigenvalues are then fitted to the level energies to determine ω, X and K. The results are tabulated in Table 10.1. The fit uncertainty is 10 cm^{-1}. This leads to an uncertainty of roughly 10% for the coefficients. (In the case of the transition state, they are 5 cm^{-1} and 5%.) Note that for a better fit, the energy ranges of HCN and HNC are divided into two regions (also shown in Table 10.1) since the coefficients may vary in a wide energy range. (The fit results confirm this possibility.)

For HCN, ω's and X's are quite the same in the lower (4194 cm^{-1} to 15809 cm^{-1} and upper (15845 cm^{-1} to 26528 cm^{-1}) energy regions. The difference lies in that K's are small for the lower energy region and more resonances, like K_{11-2}, K_{1-3-1}, K_{1-2-1} and K_{04-1} emerge in the upper energy region. This shows that in the lower region, the resonance is not eminent while in the upper region, multi-resonances are eminent by the bending motion.

For HNC, ω's in the two energy regions (8504 cm^{-1} to 16184 a cm^{-1} and 16189 cm^{-1} to 18986 cm^{-1}) are similar. As compared to HCN, the bending frequency drops from 728 (or 736) cm^{-1} to 492 (or 498) cm^{-1}. X_{22} in the upper energy region is smaller than that in the lower energy region. In the lower energy region, K_{10-2}, K_{11-2}, K_{1-2-1}, K_{1-3-1} are eminent. In the upper region, besides these terms, K_{03-1} and K_{04-1} are not small. All these show that the bending motion is full of multiple resonances.

For the transition state (energy range is from 18687 cm^{-1} to 21827 cm^{-1}), the bending frequency is only 258 cm^{-1}. Since n_2 is very large (40 to 70), the fitted K's are small. However, this does not mean that the resonances are small. Instead, they are significant. K_{1k_2-1} ($k_2 = -2, -3, -4$), K_{12-2}, K_{13-2} are of the same order. This is different from the HNC case where there are only two significant K's. For this state, multiple resonances are evident.

Besides, X_{12} and X_{23} for the upper energy region of HNC are similar and are roughly larger than those of the transition state. For the transition state and HNC in the upper energy region, X_{22} is smaller than X_{11} and X_{33}. This deserves notice though this holds also in HCN.

Shown in Fig. 10.2 are the levels of HCN, HNC and the transition state. (Note that all level energies in Table 10.1 and Fig. 10.2 are with respect to the HCN potential bottom.) The levels are up to 30,000 cm^{-1}. The location of the transition point is at $\theta = 67^0$ and 12179 cm^{-1} above the HCN potential well bottom. θ is the angle between the H–C and C–N bonds.

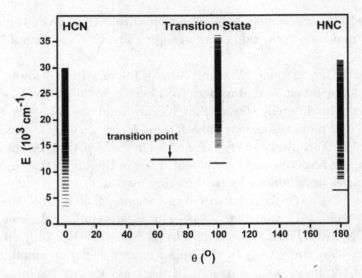

Fig. 10.2   The level diagrams of HCN, HNC and the transition state with respect to the HCN potential bottom. The longer segments are the potential minima of the respective species. The transition point is also designated.

10.3   Lyapunov exponent analysis

For a nonlinear system, the Lyapunov exponent is a key parameter to characterize its degree of chaos. For its calculation, 200 Lyapunov exponents originating from randomly chosen points in the phase space corresponding to a level are calculated and averaged. The averaged Lyapunov exponents, $\langle \lambda \rangle$, determined for HCN, HNC and the transition state are shown in Figs 10.3(a),(d). We note that only above the transition point, are the averaged Lyapunov exponents nonzero. In other words, chaos appears only above the transition point. Another notable point is the fluctuation in the Lyapunov exponents. This implies that the degree of chaos, as the level energy increases, shows a characteristic of *band structure*. Note that the fluctuations are prominent even in HNC and the transition state as shown in Fig. 10.3(d) (though they are not so clearly displayed due to the scaling of the coordinate for $\langle \lambda \rangle$). Dashed curves are drawn therein for better visualization of the fluctuations. The Lyapunov exponents for HNC and the transition state are significantly larger

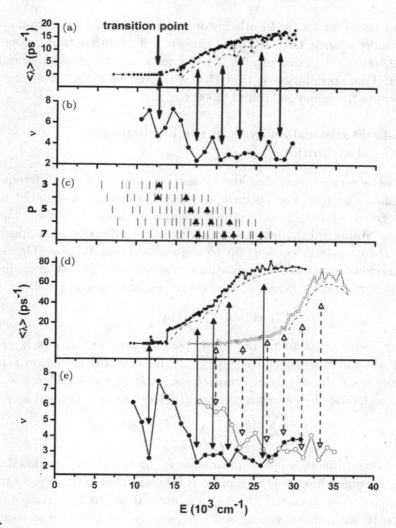

Fig. 10.3   The averaged Lyapunov exponents $\langle\lambda\rangle$ and ν for HCN are shown in (a),(b). Those for HNC and the transition state are shown in (d),(e) by solid and open notations, respectively. The dashed lines under the curves of the Lyapunov exponents are drawn for better visualization of the fluctuation of the exponents. Vertical arrows show the corresponding regions of relatively high degree of chaos, as evidenced by the Lyapunov exponent and ν. The Dixon dips, designated by ▲, for polyad numbers P = 3, 4, 5, 6 and 7 corresponding to K_{11-2} resonance of HCN are shown in (c).

than those for HCN. In other words, HNC and the transition state are more chaotic than HCN, as expected. The fluctuation is roughly of 2,500 cm^{-1} periodicity. It is not known what causes this periodicity. One speculation is that this is related to the C–N stretching since its frequency is around 2,000 cm^{-1}.

10.4 Statistical analysis of the level spacing distribution

It was demonstrated that the statistical properties are different for regular and irregular spectra. Regular spectra are associated with integrable systems and irregular spectra are associated with chaotic ones. Statistical analysis demonstrates that adjacent level spacings for these systems correspond to the Poisson and Wigner (Gaussian orthogonal ensemble) distributions, respectively. The Wigner distribution function $P(s)_{Wigner}$ by Brody[4] for level spacing s is:

$$P(s)_{Wigner} = \frac{\pi s}{2} \exp\left(-\frac{\pi}{4}s^2\right).$$

Our approach is to fit the level spacing distribution $P(s_i)_{EXP}$ to the Wigner distribution $P(s_i)_{Wigner}$ and take the deviation ν as a measure of the *departure from chaos*. That is, the smaller ν is, the more chaotic the associated spectral levels are. ν is defined as

$$\nu = \sum_{s_i} \left(P(s_i)_{EXP} - P(s_i)_{Wigner}\right)^2.$$

In calculating ν, a spectral *window* of 1000 cm^{-1} is proposed and the averaged level energy therein is taken as the ordinate for ν in its plotting. The maximal $P(s_i)_{EXP}$ is normalized to 1. s_i ranges from 0 to 10 with an increment of 0.1. s_i, expressing the level spacing, is scaled by the average of the spacings in the calculation. When s_i is close to 5, $P(s_i)_{EXP}$ is quite small, in general. The results are shown in Figs. 10.3(b),(e). To assure statistical significance, a window containing more than 10 levels is required in the calculation. Otherwise, the calculation is aborted. In the Figure, it is evident that chaos appears only above the transition point and also shows the fluctuating behavior. The correspondence between the significantly chaotic regions as evidenced by ν and the Lyapunov exponents, are shown by arrows. The regions are around 12500 cm^{-1}, 17500 cm^{-1}, 20000 cm^{-1}, 22500 cm^{-1} and 25000 cm^{-1}. The correspondence is, in general, consistent though they are based on independent algorithms and different physical viewpoints.

At this point, we want to emphasize the consistency of the calculational results by the Lyapunov and level spacing analyses of the chaotic character of the levels of HCN, HNC and the transition state, even though they possess different Hamiltonians. This assures us that the Hamiltonians adopted are appropriate.

10.5   Dixon dip analysis

As known, a resonance is associated with a pendulum motion in the dynamical phase space. Therefore, the level spacings associated with a resonance will show a dip as the level crosses over the separatrix. Shown in Fig. 10.4 are the results for the various polyad numbers (The dip first appears when P is 3) corresponding to the K_{11-2} resonance in the HCN system. K_{11-2} resonance is the most important one in the upper energy region of HCN in our analysis.

For small P, the dips are distinct. As P increases, the dips become complicated. This is due to the perturbations from other resonances so that the separatrices are interwoven together. This leads to a higher degree of chaos. Shown in Fig. 10.3(c) are the level series and the dips for HCN. It is seen that the dips do lie in the regions of high degree of chaos as evidenced by the Lyapunov exponents and ν. For HNC and the transition state, due to the complicated level structures, we did not pursue their dip analysis.

10.6   Coupling of pendulum and harmonic oscillator

We expect that the chaotic motion in the bending induced intramolecular transition can be modeled by a pendulum coupled to an harmonic oscillator. The pendulum is a mimic of the bending motion. The Hamiltonian and Hamilton's equations of motion are, respectively,

$$H = \mu^2/2 - K\cos\phi + p_x^2/2 + \omega^2 x^2/2 + k p_x \mu$$

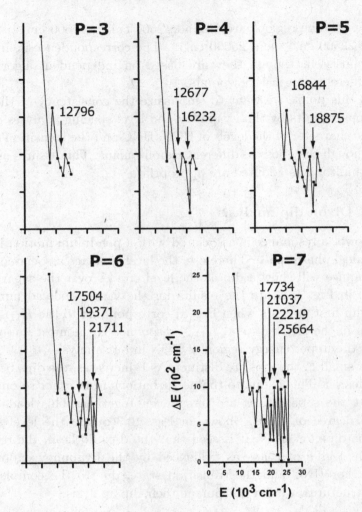

Fig. 10.4   The Dixon dips for the level series corresponding to the K_{11-2} resonance of HCN with polyad number $P = 3, 4, 5, 6$ and 7.

and

$$d\phi/dt = \mu + kp_x$$
$$d\mu/dt = -K\sin\phi$$
$$dx/dt = p_x + k\mu$$
$$dp_x/dt = -\omega^2 x.$$

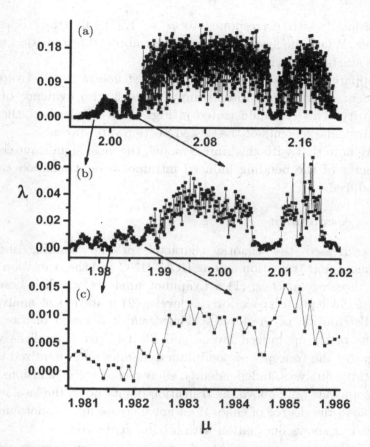

Fig. 10.5   (a) The plot of the Lyapunov exponent (λ) against the initial μ with initial $(\phi, \mu, x, p_x)_0 = (0, \mu, 1.1, 1.1)$ and parameters: $K = 1.1$, $\omega = 4.5$, $k = -0.15$. (b) and (c) are the consecutive enlargements of the portion of (a). The self-similarity structure is apparent.

In the above expressions, μ is the angular momentum of the pendulum. p_x is the linear momentum of the harmonic oscillator. ω is the angular frequency of the harmonic oscillator. K and k are the parameters for the pendulum potential and the momentum coupling strength, respectively.

Shown in Fig. 10.5(a) is the plot of the Lyapunov exponent, λ against μ with initial $(\phi, \mu, x, p_x)_0 = (0, \mu, 1.1, 1.1)$ and $K = 1.1$, $\omega = 4.5$, $k = -0.15$. For these parameters, at energies below the transition (which corresponds to $\mu = 1.98$), all trajectories are regular. When the system energy is just above the transition point, chaos starts emerging.

Apparently, as shown, there is *band structure* in the Lyapunov exponent distribution. Interestingly, there is also evidence of self-similarity. This is demonstrated in Figs. 10.5(b),(c) where the consecutive enlargements of the nested portions are shown.

We note that with this simple model, the most significant chaotic character of the bending induced intramolecular transition can be reproduced.

10.7   A comment

We addressed the chaotic character of the bending induced intramolecular transition among HCN, HNC and the transition state from three approaches: (1) a Lyapunov analysis, which is based on the classical phase space for the levels, (2) a statistical analysis of the distribution of the level spacings, which is based on the separations of the quantized levels, and (3) Dixon dip analysis which integrates the concepts of pendulum dynamics and quantized levels. The three analyses, independently, show consistent conclusions that: chaos appears only above the transition point and as the level energy increases, the degree of chaos does not increase in a monotonic way. Instead, it shows fluctuation with a band structure.

All these analyses are based on the level structure essentially. Indeed, level structure is the quantal quantity which is most easy to access experimentally. However, embedded in it, there is the classical character. This is a very important dynamical concept and constitutes the core of our works.

References

1. Bowman J M, Gazdy B, Bentley J A, Lee T J, Dateo C E. *J. Chem. Phys.*, 1993, 99: 308.
2. Bentley J A, Huang C M, Wyatt R E. *J. Chem. Phys.*, 1993, 98: 5207.
3. Ishikawa H, Field R W, Farantos S C, Joyeux M, Koput J, Beck C, Schinke R. *Annu. Rev. Phys. Chem.*, 1999, 50: 443.
4. Brody T A. *Lett. Nuovo Cimento*, 1973, 7: 482.

Appendix

Author's Publications Related to this Monograph

[1] Wu G. 'The semiclassical fixed structure of three coupled anharmonic oscillators under SU(3) algebra with Iz = 0', *Chem. Phys. Lett.*, 1991, 29: 179

[2] Wu G. 'The quasi-separable semiclassical dynamical subspace of strong Fermi resonance under SU(3) algebra', *Chem. Phys. Lett.*, 1992, 195: 115

[3] Wu G. 'Semiclassical phase space evolution of three coupled anharmonic oscillators with SU(3) symmetry partially broken', *Chem. Phys.*, 1993, 173: 1

[4] Wu G. 'The coset space representation of the algebraic molecular vibrational dynamics: The case study of SU(2) group', *Chem. Phys. Lett.*, 1993, 209: 178

[5] Wu G. 'The dynamics of energy transfer of an SU(3) algebraic vibrational system in the coset space representation', *Chem. Phys. Lett.*, 1994, 227: 682

[6] Wu G. 'An algebraic approach to the molecular electronic systems: the dynamical interpretation for the Huckel orbitals in the coset space representation', *Chem. Phys. Lett.*, 1995, 246: 413

[7] Wu G. 'The fractal character in the eigenstates of the highly excited vibrational manifolds', *Chem. Phys. Lett.*, 1995, 242: 333

[8] Wu G, 'The classical noncompact algebraic approach to the highly excited multi-mode vibrational dynamics', *Chem. Phys. Lett.*, 1996, 248: 77

[9] Wu G, Ding X. 'The SU(3) algebraic vibrational dynamics of inter-mode couplings : The case studty of H2O, CH2Br2 and CD2Br2', *Chem. Phys. Lett.*, 1996, 262: 421

[10] Wu G. 'The anharmonic effect as originated from the asymmetry of a rotor: the case study of an asymmetric rotor coupled with a simple harmonic oscillator', *Chem. Phys.*, 1997, 214: 15

[11] Wu G. 'The semiclassical dynamics of electron correlation in the doubly excited coset space', *Chem. Phys. Lett.*, 1997, 264: 398

[12] Wu G. 'Statistical interpretation for the local and normal characters of an SU(3) dynamical system and the symmetry breaking of the equivalent modes', *Chem. Phys. Lett.*, 1997, 265: 449

[13] Wu G. 'Global topological approach to highly excited vibration: a case study of H2O,Ch2Br2 and CD2Br2', *Chem. Phys. Lett.*, 1997, 270: 453

[14] Wu G. 'The classification and assignment of eigenstates of highly excited vibrational manifolds via broken constants of motion', *Chem. Phys. Lett.*, 1998, 292: 369

[15] Yu J, Li S, Wu G. 'Multifractal analysis for the coefficients of the eigenstates of highly excited vibration', *Chem. Phys. Lett.*, 1999, 301: 217

[16] Ji Z, Wu G. 'Classification of highly excited vibrational eigenstates by the nominal quantum numbers retrieved from the zero-order states: the diabatic correlation technique', *Chem. Phys. Lett.*, 1999, 311: 467

[17] Wu G. 'A classical algebraic approach to the bend motion of acetylene: the formalism by two coupled cosets', *Chem. Phys.*, 2000, 252: 315

[18] Ji Z, Wu G. 'Action localization and resonance of highly excited vibrational triatomic system', *Chem. Phys. Lett.*, 2000, 319: 45

[19] Yu J, Wu G. 'Classical characters of highly excited bend dynamics of acetylene in two coupled SU(2) coset spaces', *J. Chem. Phys.*, 2000, 113: 647

[20] Xing G, Wu G. 'Classical coset Hamiltonian for the electronic motion and its application to Anderson localization and Hammett equation', *Chinese Phys. Lett.*, 2001, 18: 157

[21] Wu G. 'The influence of the stretch modes on the classical highly excited bend motion with Darling-Dennison coupling in acetylene', *Chem. Phys.*, 2001, 269: 93

[22] Wu G. 'An alternative classical approach to the one-electron molecular orbital theory: the coset representation', *Chem. Phys. Lett.*, 2001, 343: 339

[23] Yu J, Wu G. 'The Lyapunov analysis of the highly excited bend motion of acetylene', *Chem. Phys. Lett.*, 2001, 343: 375

[24] Zheng D, Wang P, Wu G. 'The H-function for the intramolecular vibrational energy redistribution as an algebraic approach: resonances in H2O and DCN', *Chem. Phys. Lett.*, 2002, 352: 79

[25] Zheng D, Wu G. 'Chaotic motion in DCN with broken SU(2) symmetry', *Chem. Phys. Lett.*, 2002, 352: 85

[26] Zheng D, Wu G. 'Significance of the formal quantum number in the highly excited vibration of the DCN molecule', *Chinese Phys. Lett.*, 2002, 19: 466

[27] Wang P, Wu G. 'Quantization of the nonintegrable Hamiltonian by the Lyapunov analysis', *Phys. Rev. A*, 2002, 66: 022116

[28] Zheng D, Wu G. 'The study of the dynamical properties of molecular highly excited vibration', *Acta Physica Sinica*, 2002, 51: 2229

[29] Wang P, Wu G. 'Formal quantum numbers as retrieved by the diabatic correlation and their classical interpretation for the highly excited vibrational eigenstates', *Chem. Phys. Lett.*, 2003, 371: 238

[30] Zheng D, Wu G. 'Coset potential approach to the vibrational dissociation: the case study of DCN', *Chem. Phys.*, 2003, 290: 121

[31] Wang P, Wu G. 'Quantization of non-integrable Hamiltonian by periodic orbits: a case study of chaotic DCN vibration', *Chem. Phys. Lett.*, 2003, 375: 279

[32] Wang H, Wang P, Wu G. 'Dixon dip in the highly excited vibrational levels sharing a common approximate quantum number and its destruction under multiple resonances', *Chem. Phys. Lett.*, 2004, 399: 78

[33] Wang P, Wu G. 'Periodic motions due to nonlinearity: the regular vibrational dynamics of chaotic DCN', *J. Mol. Struct.*, 2005, 724: 203

[34] Wang P, Wu G. 'Quantization of non-integrable Hamiltonian by periodic orbits: a case study of chaotic Henon-Heiles system', *Acta Physica Sinica*, 2005, 54: 2545

[35] Wang P, Wu G. 'The Methods of Finding Periodic Orbit in Chaotic Systems', *Acta Physica Sinica*, 2005, 54: 3034

[36] Ou S, Wu G, Wang P. 'Chaotic motion in bending induced intramolecular transition: a case study of HCN, HNC and the transition state', *Chem. Phys. Lett.*, 2006, 432: 623

[37] Huang J, Wu G. 'Dynamical potential approach to DCO highly excited vibration', *Chem. Phys. Lett.*, 2007, 439: 231

[38] Ou S, Wu G. 'Intramolecular vibrational dynamical barrier due to extremely irrational couplings', *Chinese Phys.*, 2007, 16: 2952

[39] Ou S, Wu G. 'Correlation between chaotic dynamics and level spacings: the Lyapunov and Dixon dip approaches to highly excited vibration of DCN', *Chin. Phys. Lett.*, 2007, 24: 1841

[40] Fang C, Wu G. 'Dynamical potential approach to the dissociation of H-C bond in the HCO highly excited vibration', *Chinese Phys. B*, 2009, 18: 130

[41] Fang C, Wu G. 'Dynamical similarity in the highly excited vibrations of HCP and DCP: the dynamical potential approach', *J. Mol. Struct. THEOCHEM*, 2009, 910: 141

[42] Fang C, Wu G. 'Global dynamical analysis of vibrational manifolds of HOCl and HOBr under anharmonicity and Fermi resonance: the dynamical potential approach', *Chinese Phys. B*, 2010, 19: 010509

[43] Zhang C, Fang C, Wu G. 'Bending localization of nitrous oxide under anharmonicity and Fermi coupling: the dynamical potential approach', *Chinese Phys. B*, 2010, 19: 110513

[44] Wu G. 'The classical nonlinear properties of molecular highly excited vibration', *Scientia Sinica*, 2011, 41:410

[45] Wu G. 'A classical nonlinear approach to the molecular highly excited vibration', *J. Spectroscopy and Dynamics*, 2011, 1:3

Printed in the United States
By Bookmasters